World Ecological
Degradation

For Ben, Siang, Bill, and Barbro

World Ecological Degradation

Accumulation, Urbanization, and Deforestation 3000 B.C.–A.D. 2000

Sing C. Chew

ALTAMIRA
P R E S S

A Division of
ROWMAN & LITTLEFIELD PUBLISHERS, INC.
Walnut Creek • Lanham • New York • Oxford

ALTAMIRA PRESS
A Division of Rowman & Littlefield Publishers, Inc.
1630 North Main Street, #367
Walnut Creek, CA 94596
http://www.altamirapress.com

Rowman & Littlefield Publishers, Inc.
4720 Boston Way
Lanham, MD 20706

12 Hid's Copse Road
Cumnor Hill, Oxford OX2 9JJ, England

British Library Cataloguing in Publication Information Available

Library of Congress Cataloging-in-Publication Data

Chew, Sing C.
World ecological degradation : accumulation, urbanization, and deforestation, 3000
B.C.–2000 A.D. / Sing C. Chew.
p. cm.
Includes bibliographical references and index.
ISBN 0-7591-0030-6 (cloth : alk. paper)—ISBN 0-7591-0031-4 (pbk. : alk. paper)
1. Deforestation—History. 2. Deforestation—Environmental aspects—History. 3.
Environmental degradation—History. I. Title.

SD418 .C54 2001
333.75'137—dc21

00-066356

Printed in the United States of America

♾™ The paper used in this publication meets the minimum requirements of American
National Standard for Information Sciences—Permanence of Paper for Printed Library
Materials, ANSI/NISO Z.39.48-1992.

Contents

Figures and Tables

FIGURES

TABLES

Preface

This book explores the historical relationship between human cultures and the natural environment. Started almost seven years ago as a study on global deforestation to supplement my earlier work on capital accumulation in the timber and lumber industry, the coverage now surveys other aspects of human endeavors and their impacts on the environment. In spite of these changes, deforestation continues to be the main theme of this study primarily because of the range of ecological impacts that deforestation produces. The widening of the focus, I hope, will provide a more coherent account of the causes and consequences of ecological degradation over the course of world history than a study just on deforestation would have yielded. The coverage of this study on different geographic and cultural spaces over time enables the assessment of the relationship between Culture and Nature and to distinguish, if any, the changes in this relationship across world history. In a certain sense, it allows us to assess whether a "universality" exists in the different cultural relationships with Nature. The conclusion of this book tries to come to some preliminary assessment of this question.

In trying to cover as much time-space as possible throughout world history, certain choices had to be made about which regions and time periods to cover. These choices were made based mainly on the availability of written accounts of urbanized communities and their relations with the natural environment. As much as possible, I have tried to offer a world-historical view of connections between regions and the ecological shadows that have been cast as a consequence of these relationships. To this end, I hope others will further probe the ecologically degradative connections that I have addressed in this book.

It is always a pleasant moment to arrive at end of a project, when one gets the opportunity to thank various friends and colleagues who have provided their input. Al Bergesen, Bill Devall, Gunder Frank, Bob Marks, and George Modelski read all or most of the chapters that were written originally for different occasions and provided useful comments. Nick Kardulias and Shereen Ratnagar provided generous comments and references on the chapters that their own work covered. Others such as Bob Denemark, Chris Chase-Dunn, Jonathan Friedman, Barry Gills, Wally Goldfrank, Folke Gunther, Matthias Gross, Tom Hall, Alf Hornborg, Habib Khondker, Kristian Kristiansen, Jayant Lele, Stefan Mosemann, Jan Page, Dave Smith, Dimitris Stevis, Stephen Sanderson, Alvin So, Andy Szasz, Bill Thompson, and David Wilkinson responded to some of the themes of this project in several different meetings or as editors of volumes in which some of the earlier chapters of this book were published. Parts of this research were presented in various seminar settings at the University of Hawaii, Institute of Southeast Asian Studies (Singapore), Lund University, and the University of Bielefeld, and over the years at the Annual Conferences of the International Studies Association, the American Sociological Association, Association of Asian Studies, and the Pacific Sociological Association. I have benefited from the comments offered in these settings and from my participation in the activities and discussion of the World System History Group of the International Studies Association.

A project of this scope would not have been possible without the research support provided by Dan Blain, Randy Evans, Andrew Hund, Janine Michalek, Dan Sarabia, and Tony Silvaggio. Dan Sarabia took up the painstaking task of drawing the maps and figures for the book, and assisted in many ways during the final revisions of the manuscript. Dan Blain's unceasing efforts provided the time-series data for chapter 6, which further encouraged me that it is possible to find historical data over the long term. Matthias Gross and Stefan Mosemann spent an inordinate amount of time locating historical data and information for chapters 5 and 9, and they continue to support my efforts following their return to Germany. I am very grateful for their energies and support. Sherry Gordon and Erich Schimps (both now retired) of the university library provided generous assistance in helping me locate references and sources via interlibrary loan.

I have been fortunate to receive support from several institutions for this project. The California State University System Faculty Support Grant and the Humboldt State University Foundation awarded initial funds for the project. The Institute of Southeast Asian Studies (Singapore) provided generous research support during my stay there as a research fellow, and I am grateful to Diana Wong (deputy director) for making available its facilities. I would also like to thank Alvin So, who while during his tenure at the University of Hawaii made arrangements for a productive research stay on the Manoa campus. This book was finished at the Human Ecology Division, Lund University, Sweden, supported by a research fellowship from the Svenska Institutet, which was kindly arranged by Alf Hornborg. I am very grateful to Alf, without whom it would have taken much longer to complete this manuscript.

I would also like to thank my editor, Mitch Allen, who provided generous comments and references to the chapters and has been very supportive of this project, even though right from the start he was provided with nothing more than a long paper covering the main themes of the book. Thanks to Kelley Blewster for her careful copyediting and comments during the production process, and thanks are due also to Dorothy Bradley for seeing the book to completion.

Four persons have been very important in the writing of this book. My children, Ben and Siang, who were always willing to wait while I worked on the book, and who gave up their friends for six months to accompany me to Sweden. My dear friend, Bill Devall, who is always supportive and available to help me meet the contingencies of life. Finally, Barbro, the most enthusiastic critic of my work, who reminds me that she can do it freely because of our special relationship. She can now stop asking me in our phone conversations whether my editor has responded to the book.

Chapter One

Ecological Degradation over World History

Over world history, the relationship of human societies with Nature has been transformative. The social and ecological worlds interact with Nature's rhythms, impacting on social-economic life and vice versa. With the production of surplus as one of the key characteristics of human communities for at least the last five thousand years, the consequence has been a trajectory of numerous collisions with the natural environment as civilizations, empires, kingdoms, and nation-states seek to reproduce themselves (Chew 1997a, 1999; Ponting 1991; Perlin 1989). In most cases, the relations of Culture with Nature have been exploitative, engendered primarily to meet materialistic requirements of hierarchical systems of social organizations. The outcomes have been the loss of species diversity, polluted oceans and rivers, siltation, population losses due to flooding, health issues, and civilizational collapse (Perlin 1989; Ponting 1991; Chew 1997a, 1999).

In view of this, historically, the relations of Culture with Nature have resulted in ecological degradation (Chew 1997a; Ponting 1991; Perlin 1989; Grove 1995; Williams 1990). *What this means is that the history of civilizations, kingdoms, empires, and states is also the history of ecological degradation and crisis.* Such a historical trajectory of human "macro parasitic" activity has occurred at the systemwide structural level for at least the last five thousand years. Viewed from this long-term perspective, our present relationship with Nature has not changed significantly over time.

SOCIAL SYSTEMS REPRODUCTION: ACCUMULATION, URBANIZATION, AND POPULATION

Because of the dependency of societal systems on Nature for their reproduction, and especially the need to generate a materialistic surplus based on certain cultural consumptive levels determined by hierarchical distribution, the limits of Nature become also the limits of social reproduction. Nature as the underlying basis of the accumulation equation provides, conditions, and inhibits this process. It is the interplay between the limits of Nature and the trends and dynamics of social, political, and economic relations that ultimately defines the historical tendencies of transformation and evolution of societal systems.

On most occasions, it is a dynamic relationship whereby excessive macro parasitism by the human world of Nature's resources determines the strength and reproduction of states and civilizations, and the limits of their expansionary dynamics. It is underscored in the early civilizations whose social reproductions were contingent on economic (production of surplus) and ecological relations. These societies' reproductive and expansionary capacities were conditioned by their specific ecological surroundings coupled with the needed search/exchange in other ecological landscapes for the natural resources (timber, metals, and certain stones) they lacked and/or had already exhausted by unsustainable exploitation and accumulation. Thus, *ecological relation is as primary as the economic relation in the self-expansionary processes of societal systems.* Beyond ecological relations, natural processes and climatological rhythms also impact, and on occasion condition, the reproduction and expansionary capacities of societal systems.

In view of the above, throughout world history for at least five thousand years, the ceaseless accumulation of materialistic surplus/wealth seems ultimately self-defeating. Trade relations between developed economies and less-developed ones further exacerbate the assault on Nature. As a consequence, wider ecological shadows are cast beyond the geographic/political boundaries of the more developed economies/imperial powers where the excessive consumption and exuberant living take place.

Notwithstanding the accumulation of surplus and wealth that has severe consequences on Nature, two other world historical processes also impact on Nature: urbanization and the growth in population. To this end, they define also the Culture–Nature relations. Over world history, these latter two processes have shown exponential growth in tandem with the accumulation process.

Urbanization as a world-historical process has occurred in relation with the accumulation of surplus and wealth. By its very nature, it is highly resource consumptive in character. Large amounts of natural resources and food need to be transported to urbanized areas where manufacturing processes and the accumulation of populations have occurred. In addition to the energy inputs and other materials required for the movement of these natural resources and food products, infrastructure (such as roads, canals, terminals, ports, etc.) has to be constructed to enable such activities to be conducted. Besides the need to furnish the necessary inputs for the reproduction of life in an urbanized environment, urban areas also absorb an intensive amount of materials for the building of habitats as shelter for a dense population.

Urbanization thus is resource intensive and resource dependent on its immediate surroundings and, as well, on distant lands. The ecological shadow cast extends beyond the immediate urban confines, and perhaps extends even globally, contingent on the state of the globalization process of the world economy. Besides the intensive resource-utilization aspect of this process, urbanization also leads to the transformation of the physical landscape. Notwithstanding the loss of the aesthetic character of the landscape, the transformation of the contours of the land might lead to ecological disasters such as flooding, and perhaps to the formation of desiccated areas and swamps, depending on the nature of the ecological damage. The latter could also provide conditions for the spread of diseases. In all, the process of urbanization, depending on the nature and levels of consumption and production, generates ecological degradation when the cultural relations with Nature become extremely exploitative.

The growth in population has had an exhaustive impact on Nature. Population increase can be seen as an agent in the transformation of the landscape; throughout world history, it has borne a rather great impact (Whitmore 1993). This is especially the case when the economic and cultural lifestyles foster productive and consumptive levels that are extremely macro parasitic in nature. Where cultural patterns exhibit a regulative emphasis to restrict the growth of population, such as on island communities, environmental exhaustion does not come into play (Kirch 1997). However, when there seems to be no regulative factor or a loss of it within the cultural patterns, ecological degradation seems to result (Ponting 1991). Through isolated island communities offer a controlled method for our understanding of the relationship between Culture and Nature, it is limiting, as throughout world history there have always been numerous interactions between social systems, and in most circumstances, social systems handle their environmental resource constraints via conquests, promotion of migration (voluntarily and involuntarily), and trade exchanges.

High levels of population do impact on Nature, and the exponential growth of population over world history has generated stresses on the environment. Of course, it is not an organizing principle that all social systems with large populations necessarily impact exhaustively on Nature. Ecological degradation is a consequence of cultural lifestyles and economic organizing principles. However, over world history, it seems that when social systems reach a level of complexity and differentiation, their relationship with Nature turns degradative. Time is a key factor in ecological crisis and degradation. Nature has its rhythms and rejuvenates itself through cycles of regeneration. Ecological degradation and crises emerge when cultural transformations of Nature occur too rapidly, leaving little time for Nature to regenerate itself, or during historical epochs when the technology and practice employed modifies the intrinsic character of the constitution and diversity of Nature. This latter alteration is the most drastic, and is dependent on the cultural values of the utilization of knowledge and technologies.

DEFORESTATION OVER WORLD HISTORY

Throughout world history for at least five thousand years, the extraction of natural resources, such as wood, has been a constant feature of economic life. As such, the utilization of wood is a good proxy for our understanding of the intensity of the relationship between Culture and Nature. The utilization of the forest to meet the reproductive and expansionary dynamics of societies, civilizations, and empires has often been undertaken within and between the trade and territorial (including extraterritorial) linkages of the societies, civilizations, and empires. Besides meeting basic needs, the utilization of wood as input to the processes of production and reproduction, including the production of wood commodities, can be viewed as part and parcel of the ongoing accumulation processes occurring in societies, civilizations, and empires for the last five thousand years, especially when it is based on a cultural pattern based on conspicuous consumption. In this regard, wood has facilitated directly or indirectly the process of accumulation in world history. Trees, therefore, have fueled the socioeconomic transformation of every societal system, and for some have provided the means to become the core centers of the world system. Therefore, over world history, those states/kingdoms that are located hierarchically in more advantageous positions in the total accumulation process can extract wood products to meet their social-economic transformative needs through tributes, conquests, or trading relations.

Within all the phases of world history, socioeconomic transformations have often led to increases in population, and also in commerce. These typically have resulted in the rise of urban areas and populations with the need for housing and agricultural products. These urbanizing and population trends naturally have required the utilization and deforestation of forests to reproduce lifestyles and other social needs. The first was the need for wood in terms of basic materials for fuel, buildings, ships, and storage purposes (for example, barrels). Second, wood was required in various aspects of human activities: in the production of commodities for exchange, as a form of energy (for example, charcoal), for further extraction of other resources for production processes (such as wood required for mine shafts), and for luxuries such as scented wood products. In times of scarcity due to intensive exploitation, production processes were also relocated in order to be closer to the areas of abundant forests. Finally, trees were cut for agriculture so that the grain or livestock needs of the urbanizing population could be met and for exports to pay the taxation schema imposed by the hegemonic power. It has been stated that the advent of agriculture to meet human consumptive and exchange needs has been one of the main causes of deforestation and environmental degradation in early world history (Perlin 1989; Chew 1997a).

The above conditions often led to various pressures to move resources to feed the growing population centers of the ancient empires, civilizations, and nation-states. The normal route of transportation was shipping, which required wood for its construction (Chew 1992, 1997a; Thirgood 1981; Albion 1926). With growing levels of trade, maritime shipping increased, resulting in further wood consumption. In most circumstances, along with the increased socioeconomic activities, exuberance usually accompanied the consumption of luxuries. The latter usually led to the construction of extravagant buildings and temples, which further required timber or wood for their production and construction. To meet these demands, wood and timber were sought either through conquests or via trade relations in other areas of the world economy (Perlin 1989; Thirgood 1981; Chew 1992, 1997a, 1999; Attenborough 1987). In this regard, deforestation not only occurred within the proximity of the developed centers, but also ensued in the external arenas beyond their territorial boundaries. In the course of world history, given such conditions, degradative shadows have been cast over a vast expanse of the world economy.

CONCEPTUAL PREMISES

As expressed in the previous pages, Nature forms the basis of the materialistic reproduction of social life. Despite the significance of this, analyzing the relations and collisions of Culture with Nature has often been omitted in most attempts in the social (perhaps also the historical) sciences to decipher the patterns and trajectories of social change (see for example the works of Wallerstein 1974; Frank 1998; Mann 1986, 1993). The studies to date on long-term social and economic changes have focused overwhelmingly on the human-human side of the equation, i.e., on the social (political-economic) relations of classes, regions, civilizations, empires, and states. But are these supposedly materialist social factors and conditions sufficient to account for transformations in the *longue duree*? To be minimally materialist, the basis of social-system reproduction (in a broader context) must be viewed also through our relations with Nature.[1] By omitting the Culture–Nature relations, attempts to understand social-system transformations over the long term can only provide a partial and limiting explanation of the dynamics of historical change, and perhaps might even be misleading. An analysis of the Culture–Nature relations needs to be threaded into our overall attempts to decipher historical change over world history. As William McNeill (1990, 20–21) so appropriately states, it "deserves a place in any really satisfactory account of the past; they, too, ought to be woven into the narrative of the rise and elaboration of separate civilizations and cultures and viewed as ecumenical processes comparable in importance with the rise of a world system of economic complementarity and cultural symbiosis."

To initiate an analysis that is not anthropocentrically biased, and to obtain a more satisfactory account of the relations between Culture and Nature that generate ecological degradation over time and spatial boundaries, it is necessary to take a long-term overview. The insights emerging from such long-term and large-scale time-space considerations of Culture–Nature relations have the benefit of enabling us to track changes in patterns, trajectories, and connectivities in socioeconomic activities engendering ecological degradation that otherwise would not emerge if our analysis were based solely on a narrower time-space horizon. Of course, with such emphases, we will sacrifice perhaps certain specificities and particularities for a given period and/or geographic space. Notwithstanding this, our endeavors might also add fresh perspectives to local/regional environmental histories that might not necessarily emerge from past studies that have focused on shorter-term and micro-level changes. Additionally, it might shed some light on our understanding of whether certain social-ecological changes represent something that is specific or new to a locale/time period or reflect a more widespread (perhaps cyclical) pattern of ecological transformations. Hopefully, the reader will judge that such a course of action has been worthwhile.

HISTORICAL STRUCTURES AND PROCESSES: *PLUS ÇA CHANGE, PLUS C'EST LA MÊME CHOSE*

In the social and historical sciences, analyses of historical structures and processes over the long term and on a large scale have been undertaken quite extensively (Braudel 1972, 1981; Ekholm and Friedman 1982; Wallerstein 1974; Frank 1993; Mann 1986, 1993; Abu-Lughod 1989; Goldstone 1991; Amin 1974; Chase-Dunn and Hall 1997a; Modelski and Thompson 1996; Wilkinson 1995; Rowlands et al. 1987; Algaze 1993a). What is missing from these studies—perhaps with the exception of the work of Fernand Braudel (1972, 1981, 1982, 1984), and the writings of the Annales School, and the recent work of Chase-Dunn and Hall (1997b)—is a consideration of the relations between Culture and Nature and how these relations impact on long-term social change (Chew 1997b). Despite the absence of such deliberation, the aforementioned studies have identified historical structures and processes that are useful guides for our task ahead. The careful delineation of these historical structures, processes, and premises can assist us in sorting and providing coherence to the multitude of events, circumstances, and information that have pervaded the relations between Culture and Nature in the course of world history.

Generally, the above studies identify structures and processes of long-term change via trade and production within the framework of competitive rivalries between empires, civilizations, and states leading to core–periphery relations that further defined the trajectory of world transformations. The main corpus of such works is based on or in reaction to Wallerstein's (1974) perspective that these historical structures and processes underline a European world system that had its origins in the sixteenth century, and following its emergence has shaped the course of world transformations since then. Extrapolating the tradition of the Latin American *dependistas*, Wallerstein postulates that the dynamics of world transformations are based on the transfer of resources from the rest (the periph-

ery) of the world economy to the core areas, thus contributing to the growth of
the world economy. In turn, the former has suffered as the system has expanded
under the tutelage of the core zone, comprising certain European states contingent on the time period in question.

In spite of such a radical position, Wallerstein's periodization of system emergence and articulation of the nature of capitalism dovetails with contemporary received accounts of the rise of the West in terms of periodization and the distinct-
ness of capitalism as a mode of social organization and production (Wallerstein
1992b).[2] An uncomfortable outcome from such periodization of systemic emergence is the possible conclusion that Europe has led the way in terms of world
transformations since the sixteenth century A.D., and that prior to this period, the
world was organized around empires and civilizations that were not really stable
and that were inconsequential to the rise of the modern world.[3] What were also
not fully theorized for the period prior to the sixteenth century were the relationships and continuity between the historical structures and processes then and with
what was identified for the European world system.

From a non-Eurocentric viewpoint, it only makes sense that human history has
had a long and interdependent set of features and occurrences. Given this, there
was a plethora of Others before Europe which have played significant parts in the
making of the modern world. Other social-historical investigations have shown
that there were other world system(s) and transformations prior to the sixteenth-
century world economy identified by Wallerstein (Abu-Lughod 1989; Chase-
Dunn and Hall 1997a; Modelski and Thompson 1996; Wilkinson 1995; Frank and
Gills 1992; Chew 1997a; Ekholm and Friedman 1982; Algaze 1993a).[4] The evidence suggests that the structures and processes of the world system are of much
longer duration than what has been theorized by Wallerstein and others such as
the Research Working Group (1979).[5] The world prior to the sixteenth century
was just as vibrant, and there were other core regions that dominated the world
system when the West was in the periphery of the world economy (Abu-Lughod
1989; Frank 1998; Chew 1999; Modelski and Thompson 1996). Even McNeill
(1990), in a reassessment of his previous analysis of the rise of the West, has acknowledged that other regions/civilizations have played a significant part in the
making of the modern world, a fact that he had not acknowledged in his seminal
analysis of the history of rise of the West (McNeill 1962).

ACCUMULATION AND CORE–PERIPHERY RELATIONS

Despite the above differences, certain historical structures and processes do
seem to be enduring and recur over time: accumulation, core–periphery relations, and pulsations. Studies of long-term social change often assume that the
process of the accumulation of capital has been the driving force of world development. Capital, in this case, is used in a more general context to mean the

accumulation of surplus dependent on the state of technology and sociocultural practices of the period in question, rather than just the productive forces and the wage-labor-capital relationship commonly assumed by most Marxists.

Given this wider definition of the capital-accumulation process, it is important to determine the key components by which this accumulation process takes place. Trade exchange throughout world history has been identified as one of the underlying elements circumscribing the accumulation of capital. Much has been written on the question of whether trade exchange constitutes and enhances the accumulation process. In the Dobb/Sweezy debate over whether foreign trade engenders the capital-accumulation process, Dobb (1952, 1976) suggests that trade was not the prime factor in the accumulation of capital at the start of what has been termed *capitalism*. Sweezy (1976), however, argues that medieval trade was one of the prime factors leading to the transition from "feudalism" to "capitalism." In other words, trade exchange provides for the accumulation of capital and is part of the overall process.

Wallerstein (1974, 41–42), in his seminal work on the modern world-system, has argued that the exchange in preciosities between states and kingdoms in Europe and Asia during the fifteenth and sixteenth centuries could not have accounted for the creation of the European world-economy. This does not mean that Wallerstein discounts the effects of trade for the world system but rather that he disagrees with the type of commodity that is exchanged/traded. Schneider (1977) has challenged this Wallersteinian position, indicating that prestige goods did generate systemic effects in premodern times, whereby these goods provided local elites with important sources of power and stability. Extending Schneider's position, Ekholm and Friedman (1982) and Algaze (1993a) have further shown that besides preciosities, trade exchanges in terms of bulk goods, such as wood and grain, were undertaken by Mesopotamia around 3000 B.C. Other studies, such as Chase-Dunn and Hall (1991) and Chew (1997a, 1999), have suggested that trade exchanges included bulk goods (timber), staples (grains), and preciosities, and were important aspects of trade interconnections of the ancient world, with Frank (1993) underscoring the role of trade in the process of capital accumulation over the last five millennia. From this, it is clear that the production and exchange of preciosities and bulk goods reflect the processes of the accumulation of capital on a world scale. Thus, over world history,[6] trade and commodity exchanges do play a part in the overall dynamics of the world system(s) and thus in the process of capital accumulation.[7]

PULSATIONS

Given that the accumulation process forms the basis for the continued reproduction of the system, the world economy has been punctuated by pulsations of economic activities. The duration of these pulsations is of various lengths, stretching from fifty years to three hundred years over wide geographic space. There is a plethora of evidence indicating that early-modern economic history is punctuated by pulsations of economic periods of expansion and stagnation. Beyond economic pulsations, other cycles reflecting politics, technological innovations, and urbanization have also been distinguished in the course of world history (Chase-Dunn and Hall 1997a; Modelski and Thompson 1996; Bosworth 1995; Goldstein 1988; Freeman 1984). With the mapping of these noneconomic pulsations, it seems very likely that cycles of ecological degradation would likely correlate with the dynamics of accumulation, since excessive accumulation engenders degradative effects on Nature.

It is also very probable that beyond the rhythms of ecological degradation that accompany Frank's (1993) three-hundred-year economic phases of expansion and stagnation, there are ecological cycles of more than three hundred years in duration that could be termed as "long swings" and that recur only during certain moments in world history when the ecological landscape is stretched to its limits. Analysis of pollen count, indicative of deforestation (see figure 5.7), from the southern part of Germany over the long term has indicated extreme deforestation, and these periods fall within the time frames that historians have identified as dark ages (Rosch 1990, 1996, 1998). During these major historical moments, systems-qualitative transformations (deurbanization, production decreases, cultural lifestyle changes, etc.) occur at the socioeconomic and political levels that enable Nature to restore itself following long periods of ecological degradation. The former changes entail a downscaling and reduction in the cultural utilization of Nature's resources. Such moments are rare.

Among social historians, these historical moments are the "dark ages" of human civilizations. From an ecological point of view, dark ages are interesting periods in world history, for they exhibit trends and tendencies in Culture–Nature relations that differ significantly from expansionary phases. The socioeconomic patterns that emerge during these dark ages underline a relationship that veers away from the usual intensive exploitation of Nature normally characteristic of the trends and dynamics of human societal reproduction. For during such phases of world history—in which two periods of dark ages (1200 B.C.–600 B.C. and A.D. 500–A.D. 900) have been identified—all expansionary trends that we usually witness as typical reproductive features of human communities display negative trajectories and tendencies. For example, in the case of Mycenaean Greece during its "Dark Age," we find several Culture–Nature trends and patterns exhibiting tendencies that are subdued: fall in population levels, decline or loss in certain material skills, decay in the cultural aspects of life, fall in living standards and thus wealth, and loss of trading contacts within and without Greece (Snodgrass 1971; Desborough 1972).

From an anthropocentric point of view, such trends would spell disaster for human communities and social-economic progress, and hence the use of the adjective *dark* to depict these specific phases of world history. Ecocentrically speaking, dark ages should be appreciated as periods for the restoration of the ecological balance that have been disrupted by centuries of intensive human exploitation of Nature. Notwithstanding the visible impacts to socioeconomic life as a consequence of the onset of the dark age, this impact does not extend necessarily and evenly across geospatial boundaries of the system. Depending on the systemic connections of the world economy at a particular point in time, and the level of intensity of the Culture–Nature relations experienced by a given region, the extent of impact of a "Dark Age" period is uneven. For instance, from 1200 B.C. to 1000 B.C., the Mediterranean and the Near East region suffered socioeconomic distress while Central Europe underwent economic transformations. However, by 800 B.C. to 600 B.C., the former region was reviving while the latter declined. One would suspect that the conditions and impacts of the "Dark Age" occur in different phases for the core and the periphery/margins, and that the simultaneity and synchronicity of these conditions are contingent on the connectivity of the world system at a particular point in time. As human history evolves and with the increasing systemic connectivity among regions and the development of new technologies, these long swings are more systemic and impactive when they occur.

Given the above explication of the nature of dark ages and their rarity in occurrences, these periods are interesting phases in world history, for they offer an opportunity for us to understand the dynamics of the reproduction of human communities, and perhaps even of transitions (from one mode of socioeconomic organization to another). Understanding the pragmatics underlying the dark age and pinpointing such factors and patterns can serve our overall understanding of the conditions leading to the transformations of social systems, and the systemic limits and trajectory of the evolution of the world system in question.

To date, most attempts to explain transitions or transformations have been couched overwhelmingly within anthropocentric dimensions that are social, economic, or political in nature. From Marx to Wallerstein (1974), explanations for the transitions or transformations have been attributed to actual or potential trigger points that are socioeconomic and political in nature. In contradistinction to such anthropocentric approaches, the route that we are suggesting anchors itself within the dynamics of a Culture–Nature ecological framework. To put it in another way, perhaps it has not been "economy in command" that we have been led to believe for our understanding of transitions, but it has always been the interplay between the limits of Nature and trends and dynamics of social, political, and economic relations that ultimately defines the historical tendencies of transformation and the evolution of social systems. Nature as the underlying basis of material life provides, conditions, and inhibits the reproduction of social life. In other words, it is "ecology in command" that underlies the conditions and factors for the transformations of social systems, and it is only anthropocentricity that has conduced us to assume the former is the case.

If it is ecology that in the long run determines transformations/transitions of the world system, the appearance of these ecological cycles, though of differing durations, means that different regions of the world system(s) might be impacted economically, socially, politically, and in our case, ecologically, at the same time contingent on the state of connections/globalization of the world system. Attempts at underscoring such connectivity suggest a horizontal integration of economies and social systems over the *longue durée* that might better explain the course of world history and Culture–Nature relations (Frank 1998).

Such delineations of horizontal world history further reveal the characteristic of Culture–Nature relations, especially those perpetrated by the dominant core hegemony for a certain period in time on the periphery of the world system. In this regard, ecological-degradative shadows are cast. These shadows thus are a consequent of core–periphery relations beyond those ecologically degradative effects that might be generated by the periphery itself. In all, such tendencies highlight the compounding of the human impact on the natural environment that exacerbate further over time. The concatenation of these human-induced impacts (through trade, wars, etc.) of the environment with natural calamities (such as tectonic shifts, El Niño) and climatological changes produces ecological crisis. The latter can cause social-system collapse, depending on the state of the natural environment at that point in time, the cultural willingness and foresight to make changes in lifestyle and social organization, and, perhaps, the level of technology and knowledge available to address the conditions of the ecological crisis. The term *ecological crisis*, therefore, should only be utilized to refer to those periods of systemic ecological collapse when ecological variables exhibit severe stress and losses. Such ecological crises, we believe, are the dark ages. The rarity of such occurrences in the last five thousand years of world history suggests the resiliency of the ecological landscape to human assaults. The danger, however, is that when a dark age occurs, the tremendous socioeconomic upheavals that result, and the duration it takes for the ecological balance to be restored, could be catastrophic for social-system evolution.

CULTURE–NATURE RELATIONS OVER WORLD HISTORY

Given the historical circumstances in which kingdoms, empires, and civilizations reproduce themselves, Culture–Nature relations throughout world history can be understood via the interconnected relations between human systems and Nature over time and geographic space, and within the context of how these relations are conditioned by the processes of accumulation on the world scale, urbanization, and the growth in population. Within the ambit of the accumulation of capital, the historical structures of core–periphery relations punctuated by pulsative rhythms of growth and downturns have further defined the Culture–Nature relations over world history.

The accumulation of capital, the process of urbanization, and the growth in human population define in a systematic sense the trajectory of Culture–Nature relations. Within the geographic parameters of the world system, these processes circumscribe the pace and extent of ecological degradation and exhaustion. As such, their continued trajectory of exponential increases define specifically the trajectory of Culture–Nature relations in world history.

Such tendencies and trends of these historical processes (accumulation, urbanization, population) are fostered further and directed by the cultural patterns defined on the whole by the hegemonic political core powers for the given era and geographic space. The spread of these cultural patterns occurs within interconnected trading networks and manufacturing processes that are dominated by the core powers. Depending on the nature of the cultural patterns and the associated state of technological knowledge of the specific era, these patterns circumscribe the pace and extent of the ecological degradation that ensues. Regardless of the geographic space and the point in time, throughout world history, it seems that the cultural patterns generated by the core powers of any particular period have often exhibited the tendencies of utilizing the resources of Nature exhaustively. Thus, the degradation of the natural environment is "as old as the hills."

What follows is an initial attempt to trace within the confines of an evolving world economy the relations of Culture with Nature over different geographic spaces in world history. It is hoped that such a focus will perhaps shed some light on the historical patterns and trajectory of the human enterprise and its relations to the natural environment. Can we find a "universality" to this history of human practices regardless of cultural specificity? Is there a difference in cultural relations with Nature between epochs? These are the underlying questions that the following chapters will try to address. From a materialistic perspective, this exercise should locate us in a better position to understand and address the voguish environmental question of our times: the future of our planet.

NOTES

1. See, for example, recent attempts to address this within a world-systems perspective: Chew 1995a 1995b, 1997a, 1997b; Chase-Dunn and Hall 1997a; Barbosa 1993; Bunker 1985; Goldfrank et al. 1999; Burns et al. 1994; Kick et al. 1996; Timmons and Grimes 1997.

2. Unlike Frank (1995), who has even rejected the modes of production sequencing for understanding world transformation, Wallerstein continues to conceptualize world transformations within the specificity of capitalism as a mode of social organization and production.

3. See Wallerstein's (1974) comparison of the nature of the world system with world empire.

4. Despite the different positions taken on systemic continuity, see, for example, the collective efforts of the interdisciplinary world-system history group, underscoring the enduring pulsations and structures of (a) world system(s) that has (have) a much longer history (see Denemark et al. 2000).

5. Wallerstein (1995) has questioned whether it is possible to undertake such long-term comparisons.

6. Keeping in mind that there are some exceptions, such as Dobb's position (1952, 1976).

7. The issue of whether these exchanges occurred within a single overarching system, as Gills and Frank (1992) have argued, or whether they reflect intersocietal networks of interactions, as Chase-Dunn and Hall (1991) and Curtin (1982) have stated, is an issue. Thus, the question of whether there is a world system continuing to evolve over time or whether there were successive world systems needs to be deliberated further for our understanding of world accumulation. The work of Gills and Frank (1992), Chase-Dunn and Hall (1991), Wallerstein (1974, 1992b), Wilkinson (1995), and Modelski and Thompson (1996) has offered us different interpretations. Gills and Frank suggest that there is/are a single overarching system/successive civilizations that has/have been expanding over the past five thousand years, while Chase-Dunn and Hall see several successive world systems, each with a changing structure/mode of accumulation and its own set of hegemonies. According to Wilkinson's approach, a civilization/world system gradually engulfs other civilizations, whereas Modelski and Thompson write of an evolving system with a set of structures that experience structural evolution, thus generating social change. Resolution of these differences would be difficult; only further analysis of historical data can perhaps help.

Chapter Two

The Third-Millennium Bronze Age: Mesopotamia and Harappa 3000–1700 B.C.

CONNECTIONS

The reproduction of hierarchically based, complex social systems requires the production of surplus to meet the differential distribution of rewards and needs. This surplus production is often achieved with the intensification of the Culture–Nature relations within the immediate geopolitical vicinities and beyond. Complex and differentiated social systems also exhibit urbanized landscapes, with their specialized division of labor requiring further escalation of the utilization of natural resources and minerals. With these characteristic features that circumscribe the nature of complex and differentiated social systems, social-economic growth and perhaps demise is conditioned by the articulation of the level and nature of technology and knowledge, the cultural-political resiliencies and configurations, and their relationships with the natural environment over world history. The latter relationship is of interest to us if ecological relation is as primary as economic relation in the self-expansionary processes of social systems, and if ecological relations, natural processes, and climatological rhythms impact on and, on occasion, condition the reproduction and expansionary capacities of social systems. What were the Culture–Nature relations of the Bronze-Age civilizations such as Mesopotamia and Harappa in terms of the dynamics of their reproductive capacities? Attempting to explore this will move beyond most analyses and explanations based on the social, political, and economic factors that have been the basis of the understanding of the rise and demise of these early civilizations.

Coupled with this focus, considerations of the inter- and intraconnections among the various civilizations and kingdoms via trade, production processes, and cultural exchanges within the dynamics of an evolving world economy will shed further light on the systemic interrelatedness and the connected demise of these social systems. Given such directions, it is necessary to map out the socioeconomic practices underlining the dynamics of the world economy in the third millennium B.C., and how the former conditioned the relations between Culture and Nature, and, in turn, were also affected by them.

According to Gordon Childe (1957, 3), the regular use of copper and bronze in production and reproduction of human communities suggests the existence of an *organized international trade* during the Bronze Age. Childe's logic rests on the fact that the locations of the centers of urbanization/accumulation (such as in Egypt, Mesopotamia, and Harappa) were in the fertile alluvial valleys and plains devoid of metallic deposits. Therefore, to meet the economic need of large quantities of metals for production, other regions and areas had to supply the centers of urbanization/accumulation with the necessary metals. It reveals that a system of production, trade relations, and linkages was in place by the third millennium B.C., extending at least from Egypt through Central Asia, including the Persian Gulf, to northwestern India (Asthana 1993; Possehl 1982; Oppenheim 1979; Bag 1985; Tosi 1982; Gadd 1979; Lamberg-Karlovsky 1975; Ratnagar 1981, 1991, 1994; Crawford 1973; Edens 1992; Kohl 1987a; Allen 1992; Piggott 1950; Allchin 1982; Moorey 1994; Smith 1995; Fattovich 1985; Bard 1987; Crawford 1998; Rice 1994).

The economic system of production and trade exchanges had multiple core accumulation centers and their associated hinterlands. In the third millennium B.C., besides Egypt and the Namazga civilization of southern Central Asia, we know of two other core centers that were part of this system of economic production and exchange: southern Mesopotamia and the Harappan civilization of northwestern India. There was a series of connections occurring between the latter two far-flung urban centers, and also with their surrounding hinterlands, utilizing the land and the seas to execute the transactions. Trade was not the only means that these core urban centers relied on to meet the reproductive needs of their economies; at various times, military expeditions and wars were despatched to the associated peripheries to further the search for natural resources that they lacked and/or had already exhausted by unsustainable exploitation. Thus, we have the development of linkages that is determined by ecological conditions, trade routes, merchant communities, and political relations.

From this emerged a structure of core–periphery relations circumscribing the system of trade and production linkages that is characterized by a division of labor, with the core centers specializing in manufactured products and grains while the hinterland supplied the natural resources and minerals (Edens 1992; Kohl 1987a). The structure and processes of production and exchange in place by the third millennium B.C. were continuations, at least for southern Mesopotamia, from a millennium earlier (Algaze 1993a). Prior to the Bronze Age, there were already archaeological evidences of a system of economic production and exchange between southern Mesopotamia and its surrounding hinterlands to the north (Syria) and east (Iran). In an intensive study of the socioeconomic linkages during the fourth millennium B.C. of southern Mesopotamia, Algaze (1993a) alerts us to the series of expansionary strategies that Mesopotamia had ventured into to incorporate its surrounding hinterlands through the establishment of settlements and strategic outposts that were gateways to the resource-rich interiors. Such impulses were prompted by the need for natural resources (minerals, timber, etc.) for the reproduction of the urbanized communities of southern Mesopotamia, and perhaps by a natural calamity that led to the drying up of one of the channels of the Tigris or Euphrates, thus forcing the migration of some Mesopotamians to southwestern Iran (Algaze 1993a; Adams 1981; Gibson 1973). The incorporative strategies on the part of Mesopotamia during the fourth millennium underscore already the dynamic interactions between Culture and Nature, and how on certain occasions ecological relations and natural rhythms can impact on socioeconomic practices and condition the trajectory of socioeconomic transformation.

THE ECOLOGICAL LANDSCAPE

When the third millennium B.C. opened, the socioeconomic structural relations continued and intensified with the flow of preciosities and bulk goods circumscribing the developed urban centers from southern Mesopotamia to the northwestern Indus valley (see figure 2.1). This structure of economic and political

Figure 2.1 Trading Connections of Mesopotamia and Harappa

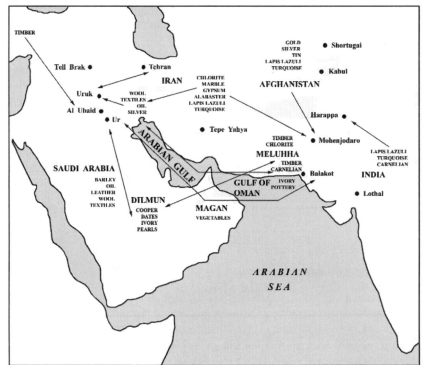

linkages covered the plains of Elam, including the Zagros ranges, the upper and central Persian Gulf, the Oman peninsula (especially the southeastern part), Persian and Pakistani Baluchistan, and the Indo-Gangetic divide. Notwithstanding certain specific ecological characteristics that existed in the river valleys, mountainous ranges, and desert conditions of this geographic expanse, the area has a common ecological feature comprising "an arid or semi-arid climate with summer droughts and an unevenly distributed and unreliable rainfall in the winter months amounting to less than 250mm per annum" (Ratnagar 1981, xiv). In addition to natural-resource extraction, this common arid ecological characteristic therefore conditioned the type of socioeconomic practice being undertaken, such as winter agriculture and the herding of sheep and goats. In riverine valleys, the natural potential for high agricultural productivity could be seen, especially in Mesopotamia and the Indus (Harappa), along with the feasibility of communication/exchanges via land, river, and sea. This potential was further increased by the use of irrigation.

Depending on the region, the mountain ranges, plateaus, and plains provided the wood and other natural resources such as metals, marble, gypsum, and alabaster for the urbanized communities in the riverine cities of the respective core centers in Mesopotamia and Harappa. Incorporative strategies by Mesopotamia and Harappa of their respective hinterlands via the establishment of outposts and settlements alongside the indigenous population facilitated the conduit of such materials. In turn, the core centers specialized in agricultural production and the manufacture of pottery, oils, wool, leather items, etc., along with wood, metal, and stone objects manufactured from imported materials for exchange. The periphery focused on natural-resource extraction and the processing of these resources for export and exchange.

Against the expanse of this structure of economic linkages, a division of labor conditioned by the ecological surroundings and the political-economic structures existed. The high productivity in agriculture, along with the specialization in manufactured products (such as textiles) facilitated by slave labor, enabled these urban centers with their associated social structures to maintain their political-economic dominance in this economic system.

MESOPOTAMIA

Culture–Nature Relations: Urbanization, Agriculture, and Manufacturing

If one were to travel from west to east in this Bronze-Age economic system of exchange, one would encounter different trading points and cultural communities reflecting various levels of development and urbanization (see figure 2.1). The urbanized communities of Mesopotamia traded in essentially three directions utilizing both land and water. The river Euphrates served the trade routes to the west and northwest (such as Syria and Anatolia), linking the eastern parts of the Mediterranean with the Gulf. To the east and northeast, the land and water transport via the Tigris tributaries linked Mesopotamia to the Iranian plateau and beyond. The southern trade routes were connected via the Gulf to Dilmun, Magan, and Meluhha.

Reproducing the materialist needs of the urbanized and stratified communities of Mesopotamia required an intensive and transformative relationship with Nature. Urbanization essentially transformed the nature of material culture and heightened its consumptive capacity, which undoubtedly led to an exchange of manufactured products for raw materials. A whole range of new materials and products, including manufacturing processes, came into being to meet these transformed material needs. They ranged from agricultural implements to cooking utensils/cutlery, furniture, textiles, transport, ornaments, and even weapons of war. What arose also were manufactories, workshops, and other industrial structures to meet the needs of urbanized communities.

In the third millennium B.C., southern Mesopotamia was composed of urbanized communities, with the principal cities of Ur, Lagash, and Susa being the main entry points for goods from the east and southeast of this political-economic system (Edens 1992; Oppenheim 1979). Canals (not only for irrigation) were built and along with natural waterways facilitated the economic exchange. Large-scale urban communities with population sizes ranging from ten thousand to more than fifty thousand were in existence and included palaces and temples (Redman 1978). Ur and Lagash had populations of thirty-seven thousand and sixty-five thousand respectively. Ur had an area of 220 acres (Childe 1964). The process of urbanization peaked early in the middle of the third millennium and reversed itself from the second millennium onwards (Adams 1981). This reversal dovetailed with the economic downturn that occurred in the second millennium B.C., starting around 1750 B.C. The scale of urbanization changed over the millennia. Urbanization reached its height by the third millennium, when more than three-quarters of southern Mesopotamia had urban areas that were over 40 hectares in size. By the second millennium (Middle Babylonian period), this proportion had dwindled to one-sixth (Adams 1981).

All this urbanization meant the intensive utilization of natural resources of the immediate surroundings and also the importation of some natural resources such as timber from distant reaches (Indus valley) of the economic system. High-quality timber from the Zagros and Taurus mountains, the Caspian area, the eastern Mediterranean, Punjab, and the Indus valley were obtained via military expeditions and trade (Rowton 1967; Willcox 1992). These imported timbers were used mainly for public buildings. Palm, poplar, and tamarisk, which were readily available in the south, were used for domestic architecture (Moorey 1994). In northern Mesopotamia, ash, oak, and elm, which were more prevalent, were utilized instead for domestic homes. Mulberry, a hardwood, was used in areas where durability was important, such as for wellheads.

The volume of timber utilization from local sources is difficult to ascertain. Potts (1977) has provided some indications from the Ur III period. Over a seven-year time horizon, from the forests of Umma, about 223 tons of wood were harvested. The forests of Umma also contributed 810 boat ribs and 59,290 pegs of wood. Besides the use of timber for housing and public buildings, wood such as the Euphrates poplar was used in the manufacture of furniture, hoe blades, plowshares, sickle handles, rungs, etc. The intensive agricultural practices for wheat and barley growing must have required a significant amount of wood to meet the needs of the grain-production process.

The relationship between Culture and Nature is underlined by the economic activities of agriculture, manufacturing, and urbanization. The ecological surroundings of southern Mesopotamia, especially its rivers, provided not only plentiful and reliable water supply but also deposits of alluvium during annual floods. Such conditions, buttressed by an extensive irrigation system of canals, dams, and reservoirs, provided high surpluses of cereals produced at low prices that were used as a basis for trade exchanges with other, agriculturally poorer regions lying east and southeast of Mesopotamia (Leemans 1960, 115; Moorey 1994).

The agricultural practices conditioned by the ecological conditions of semi-aridity required the construction of an extensive irrigation system. Initially, simple gravity-fed irrigation systems were utilized. With irrigation, fresh regions were opened up for highly productive agricultural exploitation as early as the fifth millennium B.C. Control of the flooding of the Euphrates and the Tigris was done through the construction of dams and channels. Extensive resources were used to further agricultural production. For example, the main irrigation canals were lined with burnt bricks and the joints sealed with asphalt. By 1800 B.C. the irrigation system had expanded to about ten thousand square miles (Carter and Dale 1974).

Crops grown were of several grain varieties, including other plant species for oil production and for textile manufacture. This activity thus transformed the landscape. The extensive use of irrigation required that the canals remained clear. However, Mesopotamia suffered from siltation clogging up its irrigation canals. The cause of this siltation had social roots in the extensive deforestation of the northern forests to meet Mesopotamia's fuel needs for manufacturing (copper smelting), urban social consumption, and building materials such as roof beams, posts, rods, planks, and boards (Willcox 1992; Rowton 1967). Agricultural implements such as hoes, plowshares, and sickle handles were also made out of wood. Wood was also required for shipbuilding. By the second millennium, with shortages in wood supply, ceramic materials replaced wood in agricultural implements such as sickles.

Overgrazing of the land to produce wool for textiles further exacerbated the siltation problem. Records of wool production from the Royal Wool Office when extrapolated indicate a total of 2,350,000 animals (Jacobsen 1970; Adams 1981). The sheep not only grazed on vegetation; barley was also fed to them, which required cultivation and irrigation, leading to the further transformation of the landscape.

With deforestation and overgrazing, erosion proceeded to strip the topsoil, which poured into the streams and rivers and was carried hundreds of miles downstream into canals and channels (Christensen 1993). The history of Mesopotamian agriculture is one replete with the struggle of having to keep the irrigation canals clear of silt. The more water diverted for irrigation, the slower the speed of the water's flow. Such a condition fostered a higher rate of sediment deposition in the canals. Therefore, canals and other irrigation works needed to be cleaned out regularly. When they were neglected as a consequence of social turmoil, agricultural production was impacted.

A further threat to Mesopotamian agriculture was the high proportion of dissolved salts contained in the water of the rivers descending from Anatolia and the Zagros, which were then deposited into the alluvium (Jacobsen and Adams 1958; Oates and Oates 1976; Potts 1997). The intense summer heat, with temperatures of over 120 degrees Fahrenheit, and the flatness of the land resulted in poor drainage. This, in turn, brought the salt to the surface of the fields, which in the long run had a major impact on agricultural productivity, as the sodium ions tended to be absorbed by colloidal clay particles, leaving the resultant structure impermeable to water (Oates and Oates 1976; Hughes 1975; Jacobsen and Adams 1958; Christensen 1993). High salt concentrations obstruct germination and impede plant absorption of water and nutrients.

This history of intensive utilization of natural resources (such as wood) impacted on the agricultural economic productivity, and thus determined the trajectory of socioeconomic transformation. The exploitation not only was restricted to within the immediate hinterland (up to Lebanon) but extended as far east as the Harappan civilization. Thus, the degradation of the land and exploitation of natural resources by the Mesopotamians were also transferred to other parts of the economic system.

Core–Periphery Relations

The intensive consumption of natural resources by core urbanized centers such as Mesopotamia to meet its reproductive needs not only impacts on its immediate ecological landscape but extends beyond its territorial boundaries. The set of ecological relations that resulted from such transformations was not restricted to the immediate surroundings of these urbanized communities but was extended to their hinterland areas. As a result, the resource needs of the core centers (such as Mesopotamia and Harappa) and the resultant consequences such as ecological degradation and transformation of the landscape were also transmitted ("exported") throughout the other areas of the system. In other words, an "ecological degradative shadow" was cast.

The core's ability to direct and determine the trading arrangements was a consequence of its higher efficiency and advancement in agricultural production, financial transactions and arrangements, and manufacturing. Notwithstanding these factors, its hierarchically stratified social structures exhibited a cultural lifestyle (*haute culture*) that was conveyed to the periphery through colonization and incorporative strategies, with the subsequent adoption of these lifestyle habits by the peripheral indigenous elites, which in turn further facilitated the consumption of commodities produced by the core. In all, high consumption meant transformation of the landscape with its subsequent degradation. Such trends and tendencies were exhibited by southern Mesopotamia during the Bronze Age.

Social Structure and Exuberant Consumption

The network of trading relations that existed during the Bronze Age conduces a social structure that reflects the type of commerce pursued. The different kinds of exports and imports suggest a social structure that is hierarchically differentiated. The high-yielding, irrigated agricultural system made possible the freeing up of 10 percent of the population from agricultural work (Curtin 1982; Diakonoff 1991a). Thus, we have other classes/elites taking advantage of the surplus generated: priests, temple officials, and state officials, including the military. Free persons and slaves[1] performed the agricultural tasks, the former maintaining, building, and cleaning the irrigation canals, dams, and reservoirs while on conscription duty. Ownership of agricultural land was divided among private individuals/families, the temples, and the palace (Diakonoff 1991a). Land under state ownership was cultivated, and a portion of the produce was provided as remuneration to different classes of state personnel (Ellis 1976). During the Old Babylonian period, merchants (*Tamkaru*) were also a key element in the political economy of Mesopotamia. They handled and financed the import and export of agricultural and manufactured products. This form of private financing was not the only source of capital; the temple system also financed the import and export trade.[2]

Such a pattern of social stratification produced a cultural lifestyle that was highly consumptive not only of various commodities, but of luxuries. The surplus generated based on a highly productive agricultural sector and a developed manufacturing base met the needs of those on the higher end of the social ladder and provided for the exchange of the necessary natural resources with the periphery to reproduce the political economy of third millennium southern Mesopotamia. The impact of such development for Culture–Nature relations entails the continuous pressure on the landscape to meet the exuberant needs of the urbanized communities.

Besides the amassing of capital, the consumption of luxuries and commodities by the elites and those at the lower end of the social ladder in southern Mesopotamia engendered a lifestyle that not only increased the intensity of the utilization of the natural environment, but also was transmitted to the peripheral areas. The local elites in the periphery, including the transplanted Mesopotamians, benefited economically from such trading arrangements that further facilitated the economic exchange process by adoption of the Mesopotamian lifestyles that were then trickled down to the masses. As a result, these hinterland areas required Mesopotamian products, such as oils and grains, not only for elite consumption but also for mass consumption (see for ex. Lamberg-Karlovsky 1975). This is reflected in Mesopotamia's trade relations with its eastern periphery. Manufactured goods and grains from Mesopotamia were exchanged for chlorite found at Tepe Yahya on the Iranian plateau. Yahya artisans worked at a set production rate and were paid by middlemen merchants with goods. In return, these middlemen, who functioned between the artisans and the local Yahya elites, benefited from the high profit of the chlorite sold to high-demand centers in Mesopotamia, as evidenced by the collection of high-priced items found in some third-millennium tombs in the area (Lamberg-Karlovsky 1975). The adoption of Mesopotamian lifestyles can be seen as early as the fourth millennium B.C. with the discovery of high-priced objects and pottery (Algaze 1993a).

Trade and Political Expansion

Given the structure and processes of the exchange framework, Mesopotamia's scale of economic production (of textiles, oils, leather, and other objects) with its high generation of agricultural surplus facilitated the exchange with other areas of the economic system (Adams 1981). The raw materials for the manufactured items were obtained at relatively low prices from other parts of the economic system. It was a case of "unequal exchange" (Allen 1992; Kohl 1978; Algaze 1989).[3] Imports from the north and from the eastern parts of the system (Iran and beyond)—such as metals, timber, and other items, including ivory, pearls, conch, and beads—satisfied elite and mass consumption. Natural resources and agricultural products such as timber, copper, ivory, lapis lazuli, dates, and other metals from the Gulf and the southeastern (Dilmun, Magan) parts of the system arrived at the docks of southern Mesopotamia. In return, grains such as barley, wool, textiles, and other manufactured items provided the exchange products. Silver obtained through trade (from Elam and east-central Anatolia), plunder, or tribute provided another capital source to pay for the imports (Leemans 1960, 132). The exports from Mesopotamia were usually shipped in high-prowed ships built of long giant beams tied together and coated with natural bitumen, with sails made of reed mats. These ships would range as far as Meluhha, located near the mouth of the Indus River (Diakonoff 1991a).

Transformation of distant landscape by the Mesopotamians through trade was not the only modality that occurred. Like ancient Egypt (Smith 1995), trade exchanges were facilitated by political expansion/imperialism from as early as the Uruk period (4000 B.C.) (Algaze 1993a; Diakonoff 1991b).[4] Outposts and settlements were established in the hinterland areas to control crucial nodes of trading routes along the Syro-Mesopotamian plains and in Iran. They were planted among native polities controlling communication and trade in the northern plains of the Zagros and Iran. The Zagros ranges, with their bountiful resources in timber, marble, gypsum, alabaster, limestone and other metals, were a major point of focus (see figure 2.1).

East and south of the Zagros lay Elam, which was endowed with bountiful natural resources such as timber, precious metals, and stone. In Elam, Mesopotamian merchants exchanged cattle, precious metals, oils, and timber (for shipbuilding) for wool, barley, tin, and silver. These relations were also punctuated with frequent military expeditions and wars, prompted by the Mesopotamian search for raw materials (timber, stone, metals, etc.). The Elamites sought to capture the wealth (copper, gold, slaves, etc.) of Mesopotamia. The paucity of archaeological evidence makes it difficult to assess Elam's position within this economic system. The types of resources and materials it had available for exchange suggest that its economy was based on resource extraction. It might therefore not have been as economically transformed as southern Mesopotamia. Elam, at times, was politically annexed to Mesopotamia during the third millennium.

Lamberg-Karlovsky (1986), however, presents a different view of Elam's standing in relation to Mesopotamia during certain periods of the third millennium. Elam achieved political unity by the third millennium, when there were three ruling houses: the kings of Awan, Simashki, and the "Grand Regents." Textual sources show that Awan held a position of supremacy over Mesopotamia around 2250 B.C. However, by 2325 B.C., Sargon of Akkad conquered Elam.

The imperialistic ventures of southern Mesopotamia can be divided into conjunctures (Lamberg-Karlovsky 1986). Starting in 3300 B.C., the Sumerians, with their centralized administration, started a series of colonizing missions on their periphery. Archaeological evidence indicates that a material culture identical to that of the late Uruk Sumerians developed in northern Syria, in the Zagros mountains of northwestern Iran, and in southwestern Iran. This conjuncture represented an initial attempt of the city-states of Mesopotamia to colonize the fewer bureaucratically integrated societies of the region. It was an expansionist policy that lasted for at least two centuries (Lamberg-Karlovsky 1986). The second conjuncture, starting around 2900 B.C., saw the Elamites, a regional competitor to the Sumerians, pursuing an imperialist-expansionist policy in colonizing regions of the Iranian plateau. During the final conjuncture, from 2400 B.C. onwards, expansionist Mesopotamian policies linked the Gulf and beyond.

To the east of Elam on the Iranian plateau were various communities (Tepe Yahya, Sialk, etc.) that produced the valuable metals and stones consumed by the core center. The landscape of the area was rich in minerals and stones. The indigenous community, along with Mesopotamian settlements, directed their energies toward the extraction and processing of minerals for export to the urbanized communities of southern Mesopotamia.

One of the main trading nexus of this economic system was the Persian Gulf (see figure 2.1). Sumerian and Akkadian documents underline the extensive trade between Mesopotamia, Dilmun, and Magan. Dilmun appears to be an entrepôt for the Gulf trade for goods (timber, copper, ivory, etc.) coming from the east as far away as Meluhha and Harappa, and from western centers such as Egypt, Mesopotamia, and the Mediterranean. Besides the transshipment of copper from

Oman, and timber, lapis lazuli, and other metals from eastern locales such as Meluhha, Dilmun also was a producer of dates and pearls for Mesopotamia (Saggs 1962; Ratnagar 1981; Crawford 1998; Rice 1994). After 1800 B.C., Dilmun's importance as an entrepôt declined, though it continued to export agricultural products, such as dates, to Mesopotamia. According to Saggs (1962) and Crawford (1998), this demise was a consequence of the decline in production areas such as Baluchistan and the Indus valley, which were the sources for some of the goods that Dilmun received for re-export to Mesopotamia.

Magan, too, had extensive trade relations with Mesopotamia. These relations lasted until 2000 B.C., when Dilmun seemed to have replaced Magan in trade relations in the southeastern part of the economic system, stretching from Mesopotamia to the Indus valley. Following this, Magan concentrated on its trading relations with the Indus valley. The extent of such connection is reflected in its utilization of Harappan weights and measures that have been excavated (Crawford 1998). Magan, like Dilmun, was also one of the entrepôt sites on the Gulf. Large quantities of barley, garments, wool, leather objects, and oil from Mesopotamia were exchanged for copper, precious stones, carnelian, ivory, and vegetables from Magan. The carnelian and ivory were transshipment products from as far away as the Indus valley.

The growth in the Gulf trade reached a peak between 2000 and 1750 B.C. and thus enhanced the local elites in Dilmun and Magan, who controlled this trade in terms of sourcing and producing goods for exchange with Mesopotamia. This can be seen by the expansion and urbanization process in Dilmun and Magan, where temples and tower buildings were erected (Rice 1994). For Dilmun, population levels of 170,000 have been estimated from the number of buried mounds (Rice 1994). The increasing urbanization in these places meant that the imported cereals amplified the socioeconomic position of the local elites, as they controlled the grain imports. In terms of volume of exports, such measures are not easily obtained. Ratnagar (1994), commenting on the scale of transaction for this political/economic system, writes of 18,000 kilograms of copper going through Dilmun (current-day Bahrain), one of the principal entrepôt centers south of Mesopotamia, and 2,380 *gur* (1 *gur* equals 180 kilograms of grain) of barley going to Magan, located on the southeastern part of the Gulf. Edens (1992) has referred to 18 metric tons of copper coming through Mesopotamia by the early second millennium and 2 metric tons of wool entering the system at one time. In terms of grain, volumes as high as 714,000 liters of barley were exported from Mesopotamia at one time (Leemans 1960).

Such a scale of exports and imports suggests that the goods and materials exported and imported were both luxuries and necessities in nature, with the latter targeted for mass consumption. In some cases, a luxury became a necessity once it became integrated into the political economy of consumption. For example, copper during the pre-Sargonid period was mostly imported as a luxury; with further accumulation by the post-Sargonid period it developed into a necessity when it was used widely for the manufacture of farm implements and other objects of mass consumption. This indicates a transformation from symbolic capital to the amassing of economic capital, the former referring to copper being used in objets d'art, vessels, and personal status items of the elites. The trade decline started after 1750 B.C., and description of the trade disappeared from cuneiform sources. It was revived again towards the later part of the second millennium. Notwithstanding the overall economic stagnation that occurred throughout the system during the period between 1750 and 1600 or 1500 B.C., the decline in trade volume in the Gulf during the early period of the second millennium coincided with the decline in trade of the other core center located in the easternmost part of this economic system: Harappa. It seems that the trading volume never revived, and trading activities shifted away from Dilmun and Magan and became concentrated more to the north and west of Mesopotamia with the importation of needed natural resources such as copper, precious stones, and timber.

Core–periphery relations are very degradative on the environment, especially when the periphery is providing a vast amount of resources to meet the high consumptive needs of the core, with its highly urbanized population and exuberant lifestyles. This is exacerbated further with the adoption by the peripheral elites of lifestyles and ideologies that mirror those of the core. Notwithstanding such transmutation in the cultural arena in the periphery, from an ecological point of view, the transformation of a luxury item to a basic staple in both the core and periphery further exacerbates the utilization of natural resources and the physical landscape. In the long run, the increasing pace in the accumulation of capital, fostered further in the core by an urbanized population that enjoyed an exuberant lifestyle, and duplicated to a certain extent in the periphery, led to severe degradation of the landscape. These Culture–Nature relationships were systemic—and, in fact, transhistorical in nature—during the Bronze Age. As we journey eastwards, we see that these trends and tendencies were replicated in the easternmost part of this world system.

HARAPPA

East of Mesopotamia and Elam, the predominance of Harappa,[5] the other core center in trade relations, can be observed (see figure 2.1). At the height of its expansionary phase, there were about a hundred cities spreading northwest in a chain along the banks of the Indus, indicating a high degree of urbanization. Comparable in socioeconomic transformations to southern Mesopotamia during the Bronze Age, the Harappan civilization was the core center on the eastern part of the economic system.

Highly urbanized, its population concentrated in the riverine valleys of the Indus, this ancient civilization had a set of Culture–Nature relations that duplicate what we have witnessed in southern Mesopotamia during the third millennium B.C. Connected by trade with the rest of the world of the Bronze Age, which spanned the Gulf, Central Asia, Mesopotamia, and as far as southeastern Europe, the Harappan merchants were active in moving the resources of their hinterland areas to world markets. Ports of entry and departure were sited along the coastal area; archaeological evidence has unearthed such ports, for example, Lothal.

The Harappan architectural designs and urban planning indicate a high degree of energy consumption to reproduce the predominant lifestyle. Burnt bricks, used extensively in southern Mesopotamia for building canals and ziggurats, were also the basis for building and home construction in the Indus valley. The topography of the landscape and the siting of the cities further necessitated the need for burnt bricks to be used in building to withstand the flooding by the Indus. On one hand, this flooding provided a bountiful crop in terms of agricultural production. On the other hand, it necessitated a tremendous amount of energy to manufacture the burnt bricks necessary for construction to prevent damage to homes and other structures during annual floods. Wood was the basic source of energy. High-energy consumption, to meet the reproductive needs of the urbanized population, coupled with the export of timber, to meet the demands of those distant lands such as Mesopotamia, led to extreme deforestation of the Harappan landscape and its hinterland. The Culture–Nature relations that have been observed in southern Mesopotamia were repeated here in the Indus valley, albeit within the particular nature of the ecological landscape.

Culture–Nature Relations:
Urbanization, Agriculture, and Manufacturing

In Harappa, we find a parallel to the urbanization and accumulation processes of Mesopotamia. The extent of the Harappan influence/authority reached as far west as the modern border between Pakistan and Iran, extended as far north as the foothills of the Himalayas, and stretched southwards along the west coast of India as far as the Gulf of Cambay. As in Mesopotamia, we witness again the natural potential for food production in the alluvial plains, where a large number of Harappan sites have been excavated. The urban centers were located in the river valleys of the Indus and its tributaries. The Harappan influence spread over 1.3 million square kilometers (Agrawal and Sood 1982). Its two main cities, Harappa and Mohenjo-daro, were each over 120 hectares in area, with populations around 2500 B.C. of 37,155 and 41,000 respectively (Ratnagar 1981; Allchin 1982; Bag 1985; Marshall 1931).

The degree of urbanization and architectural development can be seen in the spatial design of the cities and towns and in the architectural contours of the buildings and homes. Drainage and sewer systems were integrated into the urban design. Some urban centers were constructed according to a general plan, with streets running straight and parallel to one another and intersecting at right angles (Bag 1985). Some of the cities (Mohenjo-daro, Harappa, Ganweriwala,

Banawali, Kalibangan, Mitathal, Rakhi Garhi, Sutkagen-dor, Desalpur, Kotara Juni Karan, Dholavira, Surkotada, and Lothal) were surrounded by walls, with a citadel towering over the urban complex (Lal 1993). The large cities contained palaces, as well as granaries (measuring fifteen by six meters) to hold many hundred metric tons of grain. One great bath was also found at Mohenjo-daro, measuring twelve meters by seven meters and nearly three meters deep. Burnt or fired bricks, along with sun-dried ones, were used in enormous quantities for construction. Large amounts of fuel resources such as wood were utilized in the manufacture of such bricks (Allchin 1982; Piggott 1950; Lal 1993; Wheeler 1968). In addition, timber was used for the flat roofs of the houses, and beams with spans as large as four meters were utilized. Wood was also used for the structural framing or lacing of the brick work (Allchin 1982; Lal 1993).

Buildings and houses were connected by a sewage system; many houses had settling pits for waste water, which were directed out of the city by brick-lined channels. With the exception of the streets in Kalibangan and Lothal, every street of the urban areas was supplied with one or two drains made of burnt bricks, fine gypsum, and sand cement (Bag 1985). This level of urban transformation suggests a high consumption of natural resources, both from the immediate surroundings and from the hinterland areas, to sustain the continued development and reproduction of urban life. In addition, natural resources were also required to satisfy export markets in Dilmun and Mesopotamia (Possehl and Raval 1989).

The region in which the Harappan communities were located had the arid-zone characteristics found in other parts of this vast economic system that extended to Mesopotamia. Rainfall occurred only during winter, thus allowing for only a winter crop. Barley, wheat, cotton, and sesame were grown on the rich alluvium soil of the river valleys. Gravity-flow irrigation employing the *shaduf* lift was used in agricultural cultivation (Leshnik 1973). We can assume that there was surplus production, as the urban centers contained central granaries for storage and perhaps redistribution (Piggott 1950; Leshnik 1973).[6] Items manufactured for consumption and trade included textiles, beads, pottery, and copper and bronze objects including axes, knives, spearheads, and other tools, jewelry, toys, carvings, seals, and weights. These were made in the urban centers, while specific urban areas (lower towns) were designated for industrial production (Vidale 1989; Kenoyer 1997a). Evidence of this division of labor lies in the location of furnaces and housing for the laborers. The raw materials for these industries were chiefly copper, steatite, agate, and carnelian. Kilns of the open-firing and closed-firing types were utilized (Bag 1985). The heating of the carnelian, copper casting, and steatite glazing must have required a high volume of fuel, obtained from the immediate surroundings or from the hinterland; the heating of steatite for hardening purposes, for example, required temperature ranges from nine hundred to one thousand degrees centigrade (Kenoyer 1997a).

Such a scale of economic activity undoubtedly suggests a social structure that was stratified. Lal (1993) has proposed that three classes existed in Harappan society: priestly/king, agricultural cum merchant, and labor. However, this simplified listing might not represent the historical reality. It is highly possible, given the level of urbanism, that a state structure would have existed with its complement of bureaucrats. Included in this complexity we would also find the architects, planners, and sanitation engineers responsible for the maintenance of the physical infrastructure of the cities. We need to consider, as well, the skilled labor required for manufacturing the specialized carnelian and copper/bronze objects (Kenoyer 1989). In short, Harappan society was very much a stratified system with a division of labor and perhaps even a differentiation along ritual segregation (Kenoyer 1989).

Archaeological evidence suggests an elaborate lifestyle commensurate with that existing in southern Mesopotamia during the Bronze Age. Figurines and statues unearthed suggest conspicuous consumption by the elites. Archaeological findings of Harappan seals (indicating a receipt of goods) have indicated the presence of Harappan merchants in Mesopotamia by the end of the third millennium (Tosi 1982; Gadd 1932; Mackay 1931; Lamberg-Karlovsky 1979; Ratnagar 1981; Piggott 1950). The various types of jewelry found also reflect a class structure imbued with luxurious tastes.

Naturally, of course, as with Mesopotamia, these lifestyle trends were replicated in the periphery/hinterland of the Harappan civilization, where the elites also adopted such means to differentiate themselves from the masses. Trade between the Indus valley connected the lands surrounding the Gulf such as Dilmun, Magan, and Oman, where Harappan seals and weights have been unearthed. The presence of Harappan weights and merchant seals confirms an organized trading system administered by state structures and private merchants. These trading contacts and exchanges also led to cultural influences as depicted in the various pieces of jewelry, art, and figurines uncovered in Mesopotamia and the Indus valley (Ratnagar 1981; Waddell 1925). Exuberant consumption again came into play, and in the course facilitated the process of the accumulation of capital while at the same time degrading the natural environment.

The ecological costs for the construction of houses and buildings, the manufacturing activities, and the agricultural practices to feed a highly urbanized population were not only intensive but extensive as well, for the ecological shadow cast extended to the hinterland areas. Unlike in Mesopotamia, besides the utilization of various woods for home and building construction, the Harappans' urban communities relied on tremendous amounts of burnt bricks, as indicated previously (Hughes 1975; Piggott 1950; Ratnagar 1981; Allchin 1982; Wheeler 1968). Burnt bricks in standard sizes measuring from 11 by 5.5 by 2.5 inches to 24 by 16 by 4 inches were found in Mohenjo-daro, Harappa, and Dabar Kot (Piggott 1950). An estimate of five million bricks constituted the visible site of Mohenjo-daro (Fairservis 1979b). Such large-scale utilization of bricks for buildings, sewer systems, and canals meant that large quantities of wood were required.

This intensive utilization of a natural resource was exacerbated with the export of wood to the Gulf. Besides the intensive utilization of wood for construction and for export, there was also a heavy demand for the manufacturing of beads, copper, and steatite glazing. Fueling this resource need were the immediate surroundings of the Indus valley, the hinterland areas such as the northeastern Punjab (on the Siwaliks and the foothills), and the Western Ghats. As a consequence, extensive deforestation occurred in the Indus valley and the identified hinterlands (Hughes 1975; Piggott 1950; Ratnagar 1981, 1986; Wheeler 1968; Allchin 1982; Fairservis 1979b). Combined with overgrazing, the end result of this widespread deforestation included desiccation, flooding, and erosion.

Monoculture led to intensive utilization of the land and coupled with cattle grazing further exacerbated ecological degradation. With an increase in population, further pressure was added to increase land utilization. Fairservis (1979b, 87) has commented that "the growth of population, human and animal, dependent solely on a *rabi* crop, created seasonal stresses which in the end caused the abandonment of most of the region" of Sind. Ecological degradation of the landscape was widespread, and it increased in intensity as the millennium progressed. The continued socioeconomic growth of the Harappan civilization coupled with the growth in population pushed the carrying capacity of the environment to its limits.

Ecological degradation affected not only the urbanized communities but also the biodiversity of the area. The climatic conditions of most of the area encompassed by the Harappan civilization favored suitable habitat for the rhinoceros, the elephant, the water buffalo, the crocodile, and the tiger (Wheeler 1968; Piggott 1950). These species are depicted in the stamped seals that have been unearthed. However, none of the species seem to have survived in later periods, with the exception of the tiger. The loss of these species was a result of human activities (Possehl 1997).

Core–Periphery Relations

In line with the Mesopotamians, the Harappans in the reproduction of surplus for a stratified system also cast a wide ecological shadow beyond their immediate vicinity. For the process of surplus generation to continue unabated, core–periphery structural relations emerged to meet not only internal demands but external needs of the other parts of the Bronze Age world economy. As a consequence of these core–periphery relations, the natural environment in the hinterland areas was severely taxed to meet the exuberant lifestyles and tastes of the elites of Harappan society.

It appears that the trade and manufacturing mechanisms of the Harappans were highly evolved in terms of the mechanics of ensuring a constant supply of natural resources and a market for the manufactured products in the hinterland and in other parts of the economic system.[7] Bordering the western edge of Harappa were

Kulli communities, located along all the major avenues of communication of Makran. The paucity of large-scale archaeological excavations restricts the information available concerning these communities. Lying to the west of Indus valley, this area, especially the hills between Liari and Bela, provided the metals such as copper for Harappan manufacturing and trade (Ratnagar 1981). Harappan weights and seals have been found at Kulli sites, and the influence of the Harappans on the Kullis can be seen via the great similarity in pottery styles between the two cultures. It is also likely that the Kullis acted as merchant middlemen between the Harappans and the Mesopotamians (Piggott 1950). Therefore, the Kulli region formed part of the vast hinterland that supplied the natural resources to Harappa for its production processes. In this regard, Ratnagar (1991) has even suggested that the Kulli sites represent client chiefdoms.

The diversity of economic production and trade underlines the varied linkages that the Harappans sought, exchanged, and dominated within this region of the economic system of the Bronze Age. Harappan linkages extended to Baluchistan, Afghanistan, Iran, Central Asia, peninsular India, and the lands bordering the Persian Gulf right to southern Mesopotamia (Asthana 1982). Semiprecious stones such as lapis lazuli and turquoise were obtained from Afghanistan, Iran, peninsular India, and Central Asia for export to areas as far away as southern Mesopotamia. Securing these semiprecious stones, along with gold, lead, silver, copper, and tin, was sealed with the establishment of specialized Harappan outposts in these areas. Gateway settlements or outposts were embedded in a wide periphery to facilitate the flow of goods and natural resources (Asthana 1982; Algaze 1993b; Possehl and Raval 1989).

These settlements were established at locations near strategic trade routes/passes (e.g., Nausharo) or located in hinterland areas close to the natural resources/commodities (e.g., Shortugai) or near coastal areas to facilitate the maritime trade (e.g., Lothal). These settlements, some of which were fortified (such as Sutkagen-dor and Sutka-koh), provided the access points for the flow of natural resources much needed by the manufacturing economy of the Harappans. In addition, there were trading posts for the exchange of Harappan-manufactured goods, and some agricultural products.

Besides Afghanistan and Oman, northwestern India (Rajasthan and Gujerat) also provided copper to the Harappans (Asthana 1982; Lal 1997). The Harappans incorporated Gujerat into their political and economic sphere so as to ensure the constant flow of natural resources. Steatite and chlorite for the making of vases, seals, beads, and figurines were obtained from Rajasthan and northern Baluchistan (Asthana 1982). Again, Harappan settlements were established to secure the exchanges. The Harappans, like the Mesopotamians, would develop communities/outposts that would handle the exchange and process the natural resources obtained.

Carnelian, which was used in the manufacture of beads and pendants, was sought in peninsular India, where we also find Harappan sites. Timber, besides being a major export of the Harappans to Mesopotamia and its utilization for building and home construction, was sought in the Western Ghats, the Jammu Ranges, and the Panjab piedmont. Teak came from the Gir forests or from Panch Mahals, Surat, and the Dangs (Lal 1997). The Harappans even searched as far away as the Himalayas for timber, such as the deodar, which is a high-altitude tree. Collection centers on the western part of Gujerat were established to facilitate the flow of timber for consumption in the Indus valley, and for export via the Harappan coastal ports.

Some settlements, such as the port of Lothal, not only were procurement centers for natural resources but also supported manufacturing and modification activities (Possehl and Raval 1989). Thus, manufacturing was located not only at Harappan cities in the Indus valley but also in these established hinterland gateway settlements. It meant that the finished products, such as copper ingots, could then be exported via the coastal areas.

With processing of resources occurring in the peripheral towns, Harappan cities also had a division of labor in overall trade and manufacturing activities. For example, the production of beads was organized and controlled by elite groups (Kenoyer 1997a). Some towns were connected with different kinds of production activities. For example, Harappa specialized in metal tools manufacturing, while Mohenjo-daro concentrated on metal objects and textiles. The end result of all these economic activities would be a drastic transformation of the landscape, and in particular, severe deforestation, as a consequence of the high energy needs of the Harappans.

Such were the trading and production mechanisms circumscribing the exchange between Harappa and its vast hinterland. Besides these center-hinterland exchanges, there also occurred center-center exchanges during the third and second millennia B.C. As indicated previously, Harappan goods were exchanged throughout the Bronze Age economic system as far west as southern Mesopotamia. Dales (1977) has theorized a shift in trading routes of the Bronze Age economic system away from the Central Asian land routes to a maritime route by the mid-third millennium B.C., connecting the Indus valley via the Gulf to southern Mesopotamia. The trade on this maritime route started to decline among Mesopotamia, the Gulf, and the Indus valley by the mid-second millennium. From Harappa, copper, timber, conch shells, ivory, various stones, carnelian beads, and pigments were exported (Chakrabarti 1990). Some of these products, obtained by the Harappans from their hinterland, were then re-exported. From Mesopotamia, it was bulk goods and utilitarian items such as food stuffs and textiles through the Gulf that were exported.

THE DEMISE

The decline and collapse of civilizations have often been explained in the social and historical sciences via mostly anthropocentric reasons and causes. Usually, invasions by foreign armies, internal corruption, extreme exploitation by the elites and ruling classes are given as the factors leading to civilizational disintegration (Yoffee 1988). Explanations that attribute the fall to ecological and natural causes are rarer, and often not as popular. In view of our interest in exploring the Culture–Nature relations over world history, can we posit a possible explanation for the contemporaneous decline of southern Mesopotamia and Harappa within the confines of such parameters, including other factors such as natural calamities and climatological factors?

Economic links in terms of resource utilization and trade do not mean that the appearance of ecological crisis (as a consequence of intensive utilization of Nature) in one part of the economic system (for example, Mesopotamia) would translate to the appearance of ecological crisis in other parts (Dilmun, Harappa, Magan, etc.) of the system, though ecological degradation could be an outcome. Instead, economic crises of supply and demand could arise in these other parts as a consequence of the division of labor and trade linkages. Therefore, an ecological crisis in Mesopotamia could mean the lowering of agricultural output or production, and consequently a reduction in the overall supply of goods to other parts of the system and a drop in imports due to the lack of demand. Concomitantly, such reductions would impact other regions such as Dilmun and Harappa through a diminished demand for their materials and goods. What this type of dynamics suggests is that supply and demand might not necessarily be a consequence of the state of the economy or based on consumer tastes and needs. But rather, supply and demand can be linked inextricably to Culture–Nature relations. On certain occasions, therefore, what are seen as economic downturns generating the conditions for crisis and disintegration might have ecological roots.

If long-term economic downturns in the world system occur periodically, and if as Frank (1993) has argued they existed even during the Bronze Age, then ecological cycles of perhaps longer duration could also occur, as ecological degradation is an outcome of intensive economic activity (Chew 1997a). The demise of Mesopotamia and Harappa occurred within a systemic economic stagnation of the world system between 1800 or 1750 B.C. and 1600 or 1500 B.C. Given such economic circumstances, besides political-economic collapse, we also find ecological stress, natural disturbances, and climatological changes. Therefore, besides considering ecological relations as having an impact on the dynamics of the Bronze Age economic system that encompasses Mesopotamia and Harappa, climatological changes (in terms of rainfall and temperature fluctuations) and geological shifts could also impact on the reproductive capacities of these urbanized communities.

Given the above, explanations about the collapse and demise of civilizations and social systems need to go beyond parameters that are based solely on anthropocentric factors. In the case of the demise of southern Mesopotamia and Harappa, Culture–Nature relations coupled with climatological changes and tectonic shifts generated the conditions for the collapse. Furthermore, the demise occurred during an economic downturn of the system when there were wide-scale socioeconomic changes as well. Social-economic disruptions thus were the order of the times, and the collapse of these two social systems was symptomatic of large-scale changes occurring during this period of world history. What it would imply is that the downturn period in question was a time of systemic ecological stress (the result of the prior phase of economic expansion and consumptive exuberance), of socioeconomic reorganization and shifts in the locale of centers of accumulation. In fact, the demise of southern Mesopotamia and Harappa occurred during one of the systemic long swings of ecological stress and crisis that have durations longer than three hundred years. Deforestation data (Rosch 1990) from southern Germany exhibit such "long-swing" periods of ecological stress and crisis (see figure 5.7).

Culture–Nature Relations:
Climatological Changes, and Tectonic Shifts

Harappa

As indicated in the prior pages, the ecological relations of the two core urban complexes (Mesopotamia and Harappa) indicate an intensive exploitation of Nature to meet not only domestic needs but also commercial requirements in other parts of the Bronze Age economic system. The exploitation of Nature also underscores a drive to control Nature that is reflected throughout the history of urbanized communities for at least the last five thousand years (Chew 1997a). This exploitative nature in ecological relations upsets the balance in the natural environment, exposing such a relationship to the vulnerability of ecological stresses, natural climatic changes, and tectonic shifts. This vulnerability, coupled with a decline in trade, civil/social conflicts, and competitive state rivalries, engendered the conditions and factors for the demise of these urban complexes. Such were the ecological and social circumstances that these two urbanized communities faced.

As stated previously, the climatic zone in which the Harappan civilization extended was an arid one. Agrawal and Sood (1982), reviewing climatological data, have suggested that between 3000 B.C. and 1500 B.C. the weather was wetter in this zone. This period was followed by a severe aridity until the first year A.D. Ratnagar (1981) points as well to this climatic shift and has further suggested that a small-scale oscillation to drier conditions occurred between 1800 and 1500 B.C. This latter arid period is crucial, because it was during this time that the Harappan civilization proceeded to decline; this period dovetails with the period of downturn of the world system as indicated above (Bryson and Swain 1981; Lamb 1982a).

Especially for arid regions, minor shifts in terms of wetness (notwithstanding the major shift that occurred post-1500 B.C.) would spell severe ecological stress. Thus, starting from the second millennium B.C., the increasing aridity would place great stress on the Harappan civilization. The shift in rainfall pattern, coupled with increasing salinity, would have engendered severe stress on the agricultural productivity of the Harappans.

Tectonic shifts also occurred, which diverted water courses. In turn, these diversions transformed some rivers into dry river beds, further exacerbating the aridity and thus impacting on socioeconomic life. Agrawal and Sood (1982) make note of tectonic shifts during this period that diverted the course of the Satluz and the easterly rivers away from the Ghaggar, which over time died into a lake-like depression. The Ghaggar was alive until the late Harappan Period (1800 B.C.), but by the time of the Painted Grey Ware period (1000 B.C.) it had turned dry. Thus, in northern and western Rajasthan, unstable river systems impacted on socioeconomic life. Furthermore, tectonic disturbances also cut off Lothal from its feeder river and eventually the port's access to the sea.

This thesis of tectonic shifts impacting on the reproduction of socioeconomic life has also been advanced to account for the demise of the Harappans, especially Mohenjo-daro and its associated communities (Raikes 1964; Raikes and Dales 1977; Dales 1979; Sahni 1956). The suggestion is that a tectonic uplift generated a dam about ninety miles downstream from Mohenjo-daro in an area near Sehwan. This caused the normal discharge of the Indus to accumulate in a growing reservoir, which over time caused the flooding of Mohenjo-daro. Mud-brick platforms were erected to keep the city safe from the flooding waters. Embankments about seventy feet wide and over twenty-five feet high have been excavated. Archaeological excavations have indicated silt accumulation up to a vertical distance of seventy feet sandwiched between successive levels of occupations of Mohenjo-daro. The silt deposits are not of the type spread by normal flooding of fast-flowing rivers but seem to be of the kind similar to still-water conditions.[8] Dales (1979), building on this, has suggested that, enmeshed in this deteriorating condition, the weakened state was unable to send help to its inhabitants in the northern frontier when they were threatened by tribal incursions.

Beyond the arguments suggesting that climatological changes and tectonic shifts brought about the demise of the Harappans, it has been suggested that overcultivation, overgrazing, salinity, deforestation, and flooding contributed to the decline of the Harappan urban complex (Fairservis 1979a; Raikes and Dyson 1961; Wheeler 1968; Stein 1998). The flooding of the Indus would have had the effect of increasing the water table for a considerable distance on both sides of the channel. With a high salt content in the water table, and the mean temperature about ninety degrees, poor leaching led to large areas becoming unfit for cultivation over time. Along with the wearing out of the landscape through overcultivation, overgrazing, and deforestation, this would have generated severe stress on socioeconomic reproduction. It has led Fairservis (1979b, 88) to suggest that "the evidence again demonstrates a failure to come to grips fully with what must have become an increasingly acute situation—the destruction of the local ecological patterns and the consequent failure of food resources."

The identification of the above conditions as possible factors leading to the decline of the Harappans around 1700 B.C. furthers the need to consider ecological relations, natural climatological changes, and tectonic shifts as important variables impacting on the reproduction of the Harappan civilization.[9] This consideration needs to be coupled with other anthropogenic factors such as the decay of administrative structures, social conflicts, and border threats in the north and west (Dyson 1982; Piggott 1950). Notwithstanding all these anthropogenic and ecological factors that are *specific* to geographic parameters of the Harappan civilization, we need to consider the economic linkages among the Harappans and other parts of the Bronze Age economic system that also facilitated the demise of this urban complex (Lal 1997).

The decline of Harappa coincided with that of southern Mesopotamia. Trade between Harappa and Mesopotamia and the Gulf region, as we have indicated, flourished from 2500 B.C. until 1750 B.C. After this, the Gulf trade disappeared (Leemans 1960). The end of this flourishing maritime trade coincided with the time period of the demise (1700–1500 B.C.) of the Harappan civilization. The slowdown in the Gulf trade and the demise of the Harappans coincided with the downturn in the world system starting around 1800 or 1700 B.C. and lasting until 1600 or 1500 B.C. As the urbanized communities of Harappa were linked to the overarching Gulf trade and beyond, its infrastructure and surrounding hinterlands had therefore developed and specialized in the manufacture of products and natural resources for export. Thus, when its exports to the Gulf and beyond disappeared, it could no longer reproduce the accumulation process that had sustained the urban growth. This led to migration to the rural areas of the north and south. With the accumulation process slowing down, the outlying trading towns and outposts in the hinterland, devoid of prosperous support from the Harappan core, gradually merged with the countryside. Coupled with the problems of (normal and abnormal) flooding and the deteriorating ecological conditions, the decline proceeded unchecked.

The contemporaneous decline of two core centers of accumulation underlines the systemic connectedness of the Bronze Age economic system. Harappan exports were impacted by the slowdown in Gulf trade and the declining political eminence of the southern cities of Mesopotamia where the exports entered. The Mesopotamian decline also has its roots in the ecological relations that existed, along with climatological changes in terms of temperature and rainfall.

Mesopotamia

Between 3100 and 1200 B.C. the Near East experienced a drop in precipitation, and we find that wetter conditions did not return until 850 B.C. (Neumann and Sigrist 1978). This increase in aridity was followed by much warmer weather during the winter months. Warmer and drier conditions were also reported for the whole eastern Mediterranean area around 1500 B.C. (McCoy 1996). The increase in temperature, especially between 1800 and 1650 B.C., has also been confirmed for Mesopotamia by comparing harvest dates for barley during this period with a later period, that is, after 850 B.C. (Neumann and Sigrist 1978). These climatological changes, occurring during a period of economic distress of the system (1800 or 1750 B.C.–1600 or 1500 B.C.), would have further exacerbated the already strained conditions. In the agricultural sector, especially with an increase in temperature, they would have led to a rise in the evapotranspiration. For irrigated agriculture this would mean a demand for more water. The enhanced application of irrigated water had a deleterious effect on agricultural lands that possessed a salinity problem. Conditions were prevalent in southern Mesopotamia between 2400 B.C. and 1700 B.C. that led to a crisis in agricultural productivity (Jacobsen and Adams 1958; Adams 1981).

This crisis in agricultural productivity has to be understood within the wider contexts of the political-economic relations and the ecological relations of Mesopotamia. No doubt climatological changes such as warmer temperatures and decreasing rainfall are serious factors relating to agricultural productivity, but such shifts might not have been as impactive if the ecological relations of Mesopotamia had attempted to maintain a balance with Nature.

As stated previously, this was not the case. The stratified society pursued intensive socioeconomic activities to produce surplus for domestic consumption as well as for exports (in the form of grains and woolen textiles) to the Gulf and beyond. The scale of intensity required extensive deforestation, maximal utilization of agriculture, and animal husbandry. Furthermore, with a state structure requiring tax payments, etc., the farmers needed an increasing surplus to meet the reproductive needs of the system (Jacobsen and Adams 1958). Besides requiring surplus production, state direction during the Third Dynasty of Ur (2150–2000 B.C.) also concentrated on certain economic activities such as the production of wool and the development of a large-scale textile industry. This further pushed the need to increase agricultural productivity in the form of feed grains such as barley for the sheep (Adams 1981). Population increases and state initiatives to establish new towns populated by conquered peoples for the purpose of pursuing agricultural and textile manufacturing added to the range of economic practices that required heightened resource utilization (Gelb 1973). The end result of these political and economic initiatives was an intensification of agricultural production that pushed the agricultural lands to the limit.

As indicated previously, salinization was a problem in Mesopotamia from the third millennium onwards. Wheat and barley that were grown in the mid-fourth millennium in equal portions had shifted to more barley cultivation by the end of the third millennium. With the salinity problem, after 1700 B.C. the cultivation of wheat was completely given up, as barley was a more salt-tolerant grain. In addition, barley was also the preeminent feed for sheep, whose end product, wool, was a commodity required for long-distance trade. However, this change in grain cultivation did not solve the issue of soil salinity, which is reflected in the harvest yields. Between 3000 and 2350 B.C. (Early Dynastic Period), crop yields were 2,030 liters per hectare. By 2150–2000 B.C. (Third Dynasty of Ur), the yield had fallen to 1,134 liters per hectare. By 1700 B.C., crop yields had slipped to 718 liters per hectare (Jacobsen and Adams 1958). The agricultural-productivity crisis reached its nadir when cultivable land was kept in production with yields of only 370 liters per hectare (Adams 1981). With such yields, the "burden on the cultivator had become a crushing one" (Adams 1981, 152). Such desperate conditions occurred during a period of world-systemic economic stagnation—between 1800 or 1750 B.C. and 1600 or 1500 B.C.—that further exacerbated what were already stressful times.

To meet this agricultural crisis, seeding rates were increased. Between 2150 and 2000 B.C., 55.5 liters of seed were planted per hectare. This volume doubled in comparison to the period between 3000 and 2350 B.C. Furthermore, with no consideration for the repercussions, alternate-year fallowing was also violated in order to maintain the maximum amount of land available for cultivation (Gibson 1970). Fallowing is the traditional method of handling salinization. In Mesopotamia, when the land was left fallow, wild plants such as *shok* and *agul* drew moisture from the water table and dried up the subsoil. This prevented the water from rising and bringing the salts to the surface. When the land was cultivated again, the dryness of the subsoil allowed the irrigation water to leach salt from the surface and drain it below the root level. Fallowing therefore returned the land to its cultivation potential. By reducing or violating fallow times, productivity of the land was endangered.

Southern Mesopotamia never recovered from the disastrous decline in agricultural yields that accompanied the salinization process. Deurbanization was the order of the day, especially toward the end of the second millennium (Adams 1981). Urban life and culture continued on a declining scale with the population concentrating only in the major towns (Brinkman 1968). As a consequence, the cultural and political centers of power shifted from the south to the north with the rise of Babylon in the eighteenth century B.C. This explanation for the demise of southern Mesopotamia and the resultant transfer of power to the north as a consequence of ecological relations and climatological changes needs to be deliberated with the explanation for the collapse of the Third Dynasty of Ur as a consequence of the activities of tribal groups such as the Amorites and the coalitions that resulted (Yoffee 1988). When combined, these anthropogenic and ecological explanations enrich our understanding of the vectors that led to the demise of southern Mesopotamia and the consequent rise of Babylon, for as we indicated in the previous chapter, ecological relations and political-economic relations are intertwined in conditioning the transformation of regions of the world system. Furthermore, when the power base shifted northwards, the trade in metals and timber with partners in Syria and Anatolia increased. The central Mesopotamian cities thus relied more on this northern hinterland for their resources than the southern cities did. Such developments led to the strengthening of the land trade routes of the north and northeast and the demise of the maritime trade routes of the Gulf. The shift engendered the disappearance of the Gulf trade, and the repercussions were felt throughout the trading points along the Gulf up to the Indus valley.

The contemporaneous decline and demise of Harappa and southern Mesopotamia were consequences of their economic connectedness in the Bronze Age economic system and their comparable ecological relations with the natural environment. The character of these ecological relations is circumscribed by the dynamics of the process of unequal accumulation, the pace of urbanization, and the population densities that existed then. In addition to these processes, exuberant lifestyles furthered the need for an intensive utilization of ecological resources. Ultimately, the continued inability to sustain unequal accumulation, urbanization, and population while maintaining a successful balance with the natural environment leads to crisis, especially during a conjuncture when ecological stress, climatological changes, tectonic shifts, civil strife, state rivalry, border incursions, and wars punctuate the historical moment. Centuries of ecological degradation as a consequence of Culture–Nature relations have the cumulative effect of stretching the ecological system to its limits, whereby an ecological crisis results that in turn impacts on socioeconomic reproduction. It is suggestive from the long-term deforestation time series (see figure 5.7) that the demise of Harappa and Mesopotamia occurred during such a period of systemic ecological long swing (1700 B.C.–700 B.C.), whereby Nature was severely stressed, which in turn, affected socioeconomic reproduction in the long run. Systemic in nature, the ecological long swing pervaded the second-millennium world system. We also witness such conditions in the eastern part of the Mediterranean, in the semiperipheral and peripheral areas of the world system. It is to this we now turn.

NOTES

A previous version of this chapter was published as "Ecological Relations and the Decline of Civilizations in the Bronze Age World System: Mesopotamia and Harappa 2500 B.C.–1700 B.C." in *Ecology and the World-System*, edited by W. Goldfrank et al. (Westport, Conn.: Greenwood Press, 1999), pp. 87–106. Many thanks to Mitch Allen, Gunder Frank, Bob Marks, George Modelski, Shereen Ratnagar, and Andy Szasz for their comments.

1. Slaves existed throughout the third millennium B.C., and they provided part of the labor in the agricultural and manufacturing sector. By 1900 B.C., the Larsa period, we also witness the transformation of free persons into slaves as a consequence of debt or hunger or through sales of children by parents (Saggs 1962).

2. Contracts in the form of tablets were drawn up to underline the conditions of the loan and the expected rate of return; they were sealed with the cylindrical seals of the concerned parties. The tablets were then enclosed in a clay envelope also containing a duplicate copy to render falsification impossible (Saggs 1962). Interest payments varied from 33⅓ percent for barley to 20 percent for silver (Saggs 1962; Oppenheim 1979). The creditor thus provided the capital in return for a fixed rate of interest on the loan without having the

opportunity to share in the profits or losses that arose from the transactions. Saggs (1962) has indicated that this form of financing, with a fixed return to the creditor, applied only to the maritime trade, whereas terms for loans applying to the land-based caravan route were different. In the latter case, the creditor was entitled to two-thirds of the profits, with a guaranteed minimum return of 50 percent of the outlay of capital. Such a difference in terms was due to the technical skills supposedly required in the maritime trade and the special access that the seafaring merchants had to the trading centers in the network of this economic system. Unlike Polanyi (1957), others (such as Saggs 1962; Silver 1985; Powell 1992; and Lamberg-Karlovsky 1975) have argued that the commercial and financial activities as early as the third millennium B.C. in Mesopotamia fall within the interplay of market forces. Long-distance trade, differential exploitation, and responses to supply and demand characterized the commercial activities of the period (Farber 1978). Historical accounts indicate price increases in grains during the Akkadian Dynasty of 100 percent. Price increases were a consequence of wars and intensive economic activities resulting in inflation. There were also reports of price decreases as a consequence of oversupply.

3. This descriptor underlining trade exchanges is derived from the work of Emmanuel (1972). It is used descriptively to depict the trading arrangement.

4. During periods of state expansion, such as the Akkadian period, the military leader, Lugalzagesi, proclaimed sovereignty over Sumer lands to the west and east of Sumer (Lamberg-Karlovsky 1975). Similarly, Sargon of Akkad claimed control of Lebanon and the Ammanus ranges, which were rich in natural resources and trees (Saggs 1962). During the reign of Sargon II, the king of Dilmun, to underscore his vassal status, sent tribute and soldiers to assist Sargon II in his war campaigns. Dilmun had extensive trade relations with southern Mesopotamia. Akkadian imperialists twice invaded Magan, which is located to the south, and resistance from southeastern Arabia led to strengthened ties with the people of the Indus (Edens 1992).

5. The use of the word Harappa/Harappan is not to designate specifically the city of Harappa in the Indus valley during the Bronze Age. It is used to represent the urbanized communities located in the Bronze Age Indus valley that are known historically as forming the Harappan civilization.

6. Ratnagar (1986) has questioned the agricultural productivity of the Harappan region in view of the ecological conditions requiring lift irrigation vis-à-vis the agricultural productivity of southern Mesopotamia of the same period.

7. Our position differs from that of Shaffer (1982), who has indicated that the Harappan civilization was primarily dependent on internal exchange.

8. This thesis has been questioned by Lambrick (1967).

9. The dating for the decline of the Harappan civilization varies according to different Harappan scholars. Wheeler (1968) has suggested that the earliest decline dates from 3250 B.C.–2750 B.C., while Gadd (1932) has pinpointed 2350 B.C.–1700 B.C. as the period of decline based on his analysis of the seals. Ratnagar (1981) has identified the period between 2000 B.C. and 1760 B.C. as the phase of decline, while Possehl and Raval (1989) have suggested that the phase ended later, around 1000 B.C. Fairservis's (1979b) dating of 1750 B.C. as the end seems to encompass these differences, and it also dovetails with the decline of the Gulf trade, in which the Harappans participated actively.

Chapter Three

The Second-Millennium Bronze Age: Crete and Mycenaean Greece 1700 B.C.–1200 B.C.

If one follows the trail of this Bronze Age system of trade exchanges northwest from Mesopotamia, we reach the eastern Mediterranean, where by the third millennium B.C. there were communities developing along stratified lines (Kardulias 1996; Renfrew 1972). Trade connections all over the eastern Mediterranean during this period linked this region to Egypt, Syria, and Mesopotamia. Between 2500 and 2000 B.C. population growth further transformed the region (Sheratt and Sheratt 1991). Nonetheless, the Aegean continued to be in the peripheral zone of the Bronze Age world system in relation to the other core areas in Mesopotamia, Anatolia, Egypt, and northwest India.[1]

Following an expansionary trend beginning in the twentieth century B.C. and lasting until 1700 B.C., Crete, the Cyclades, and later the Greek mainland (Mycenae) were transformed into intermediary trading centers of the eastern Mediterranean. They had linkages with eastern Europe, especially Romania and the Otomani culture. Bullion and amber were exchanged between eastern Europe and Mycenae during the second millennium B.C. (Kristiansen 1993; Dickinson 1977).[2] By the middle of the millennium (circa 1600 B.C.), besides eastern Europe as the source for raw materials on which the developed production processes in the Aegean depended, the Baltic region also became a supplier of amber (Harding 1984). Increasingly, with the continued expansion of the manufacturing and production processes throughout the second millennium in the Aegean, Europe was drawn into the Aegean political economy and incorporated further into the Bronze Age world system. The river systems of Europe, such as the Saone and Rhone, provided the main transportation routes.

Within the Bronze Age system, the demise of the Persian Gulf trade by 1750 B.C. resulted further in an increase of the trade of the eastern Mediterranean with the Anatolian plateau. The collapse of the Gulf trade occurring within a conjuncture of economic stagnation enabled the rise of intermediary centers in the periphery. Crete benefited from the expansion of trading activity in the eastern

41

Mediterranean and later Mycenae. It had already been on a growth curve since the start of the second millennium, with its command-palace economies importing raw materials for its manufacturing centers and exporting its products eastwards. Such an increase in demand also benefited Mycenae, and by the mid-sixteenth century B.C. Mycenae became a dominant center of the Aegean.

Expansive growth—similar to what was discussed in the previous chapter for southern Mesopotamia, the Gulf, and Harappa in terms of expansion of trade, urbanization, and intensification of production and manufacturing processes throughout the third millennium—was repeated in the eastern Mediterranean part of the system. Except in this case, the transformations occurred a millennium later, during an economic stagnation that most often benefited and enabled economies in peripheral and semiperipheral areas to rise.[3] Again, to continue our historical journey analyzing ecological degradation across time-space dimensions, we examine a peripheral/semiperipheral area of the world economy a millennium later to distinguish the state of the ecological landscape. As a propaedeutic exercise, we will scrutinize the demarcation of the dynamics, extent, and connectivity of this part of the Bronze Age world system with its underlying process of accumulation. The outcomes of such connections and economic activities, and the consequences to the natural environment, will be revealed along with an articulation for the demise of the Aegean communities during the second millennium B.C.

CULTURE–NATURE RELATIONS IN CRETE

Urbanization and Population Growth

Crete is a mountainous island, straddled by three mountain ranges with peaks ranging from 2,000 to 2,500 meters. Much of the land surface is about 300 meters above sea level, and the island is 260 kilometers from west to east with the widest north–south point measuring about 55 kilometers. The climate is Mediterranean, with very hot summers and rigorous winters and temperatures rarely falling below eight or nine degrees centigrade. Such an ecological zone—with mountain ranges forming the backbone of the island landscape and valleys providing the alluvial plains for agricultural cultivation and animal husbandry—structured the socioeconomic practices of ancient Crete.

At the start of the second millennium B.C., in a period of economic expansion, urbanized communities developed that were centered around the palaces and villas (Renfrew 1972; Chadwick 1976; Warren 1985). Archaeologists have divided the palace-building periods into two phases. The first phase, from about 1950 B.C. to 1700 B.C., is considered the First (or Old) Palace Period. It occurred during the time when the world economy was expanding, and Crete rode along the crest of this economic expansion as a conduit for raw materials from Europe and its own agricultural products to west Asia, Egypt, Mesopotamia, and the Gulf. It was thus an exuberant time for the elites in Crete. Palace building was the order of the day. With economic expansion, ecological degradation followed. Such exuberance came to an end with the economic stagnation of the system starting from 1700 B.C. onwards, when collapse of trade and demise of civilizations occurred in other parts of the world system, as discussed in the previous chapter.

In spite of these economic and ecological circumstances, another period of construction occurred from 1700 B.C. to 1450 B.C., which is often treated as the Second (or New) Palace Period. The break from the First Palace Period also dovetails with the period of stagnation of the world economy. In certain parts of the world system, deurbanization and collapse were evident and ecological stress was pervasive; this was especially true during the conjuncture of an ecological long swing, when ecological crisis pervaded the system.

During the Second Palace Period, there was a shift from highly resource-consumptive palace building to the construction of smaller-scale structures such as villas. Could the shift to smaller-scale structures have been a consequence of the limits of natural resources and ecological stress, or was it an economic and cultural evolution? Some archaeologists who have signified the break in palace building have suggested that such a break was to depict a period of downturn and destruction of the physical and economic life in Crete; others who have focused on socioeconomic aspects have suggested that the building of villas occurred at a time when storage areas in the palaces were much reduced and the loss was made up by villas (Manning 1994). Perhaps ecological constraints and economic and cultural evolution all played a part in this transformation of the urban landscape.

With the appearance of villas, there was a shift toward decentralization of building construction and, as well, the emergence of a landed elite tied to the palaces. Villas also alleviated the space constraints of palaces and enabled the transformation of the Minoan economy towards manufacturing. The storage spaces for oils and grain were reduced, and manufacturing areas in the palaces during the Second Palace Period were increased, utilizing the space originally slated for storage in the First Palace Period (Manning 1994; Halstead 1981).

This shift further underlines the transformation of the Minoan economy and its shift to value-added manufacturing in addition to agricultural exports. The change occurred within a conjuncture when the Gulf trade was declining and new points needed to emerge to fill the gap for the demand of manufactured products following the decline of southern Mesopotamia and Harappa, which in the past had been main suppliers of the system.

The most famous of these excavated palaces, Knossos, centrally placed on the north coast of the island, is surrounded by a large fertile area to the south, the plains of Messara (see figure 3.1). To the east of Knossos lay another palace complex, Mallia, which would have likely been subjected to the control of Knossos. Further east of Mallia lay the complex Kato Zakros. South of Knossos, under Knossos's control as well, was the palace complex of Phaistos.

Figure 3.1 Bronze Age Crete

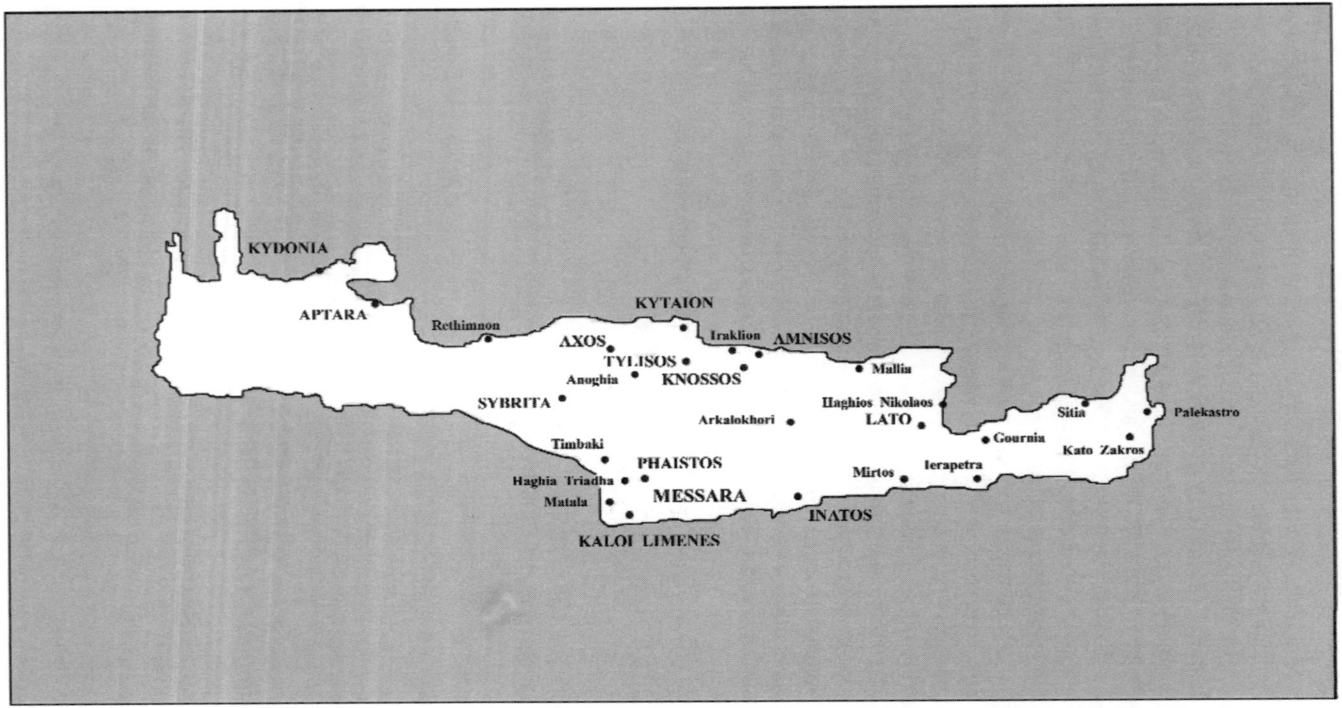

Knossos, Mallia, and Phaistos were surrounded by fertile alluvial plains for agricultural production, while Kato Zakros seems to have functioned only as a port for trade with the eastern Mediterranean. The issue of whether Knossos dominated the political economy of Crete has been widely debated. Renfrew (1972) and Chadwick (1976) have argued that it is very unlikely that Knossos controlled the whole island. Others, such as Walberg (1960) and Cadogan (1983), believe that during the Old (First) Palace Period, each palace was a single political entity. By the New (Second) Palace Period, Warren (1985) and others, such as Hood (1983), Cadogan (1984), and Weiner (1987), have suggested the dominance of Knossos in the political economy of Crete. Others, such as Cherry (1986), Bennet (1990), and Soles (1991), however, have argued that several independent palace polities existed.

Regardless of whether there was unitary political control or multiple political entities, economic specialization continued and expanded, resulting in an intensive utilization of Nature within the island and other areas of the Bronze Age system supplying natural resources to Crete. This expansion was accompanied by new settlements, comprising buildings, grand country houses, and town houses (Cadogan 1984). By the second millennium B.C., the Minoans had built paved highways, streets, aqueducts, and sewer systems to serve the urbanized communities (Carter and Dale 1974). They also built extensive drainage projects for agricultural production. The extent and sophistication of this infrastructure construction "were superior to anything found in the Mediterranean region before Roman times" (Carter and Dale 1974, 64).

According to Renfrew (1972), the number of settlements in Crete, besides the palace complexes, grew by a factor of 2.6 from the Neolithic period to the early Bronze Age, and this factor rose to 6.8 times from the Neolithic period to the late Bronze age. Settlement sizes varied. The Phaistos and Mallia palace complexes ranged from 8,400 to 9,800 square meters in the late Bronze Age, while Knossos was about 13,000–17,400 square meters (Renfrew 1972; Warren 1985). Zakros was by far the smallest of the complexes, with an area of 4,250 square meters.

In comparison to the urban settlements of Mesopotamia and Harappa, these palace-complex settlements were much smaller in scale—but not in terms of complexity of architecture and the utilization of natural resources. Measured on the Mesopotamia-Harappa urbanization scale, the palace complexes at Crete were no larger than fair-sized villages of the Early Dynastic Mesopotamia. Such a difference in scale locates the politicoeconomic position of Crete and the Aegean outside the core areas of the second-millennium Bronze Age system. If we use this measure, plus the population levels (to be discussed next), it would appear that the Aegean cannot have been part of the "core" of the system. Nevertheless, on a regional basis, Crete was a dominant economic center.

By the early Bronze Age, Crete's population was approximately 75,000 persons; it grew to 214,000 persons by the middle Bronze Age. At the end of the late Bronze Age, it reached a population of 256,000 (Renfrew 1972). Population increased throughout the Bronze Age and began to slow down from about 1600 B.C. onwards, dovetailing with the phase of stagnation of the Bronze Age cycle between 1700 and 1500 or 1400 B.C. (Frank 1993). Knossos, the largest of the palace complexes, had control over lands where the population level is estimated at about 50,000 persons. This means that by the late Bronze Age it controlled about a fifth of the population of Crete. Other settlements, such as Mallia, Mokhlos, Phaistos, Khania, Gournia, and Palaikastro, also grew to sizes similar to Knossos. Urbanization surrounded these palace complexes, where intensive settlement—for example, at Knossos—covered over seventy-five hectares. These settlements included large and impressive private homes finished with "neatly jointed ashlar masonry" (Warren 1985).

During the Old Palace Period (1930–1700 B.C.) at Mallia, there were also buildings that had cult rooms and workshops (Chadwick 1976; Renfrew 1972). The level of utilization of natural resources required to build these palace complexes and settlements can be seen in the scale and complexity of the buildings. All the palaces had at least two floors and were divided by functional requirements (Chadwick 1976; Renfrew 1972). Sections were devoted to storage of grains and oils. At Knossos, storage capacity for olive oil alone is estimated to have been around 246,000 liters. This volume occupied only a third of the storage area at the palace of Knossos. There were also areas allocated for industrial processing of oils and wine. Swimming pools and running water were available in the living quarters.

Given the above socioeconomic practices, the Culture–Nature relations were quite intensive and expansive. All the land areas in Crete were devoted to meeting the reproduction of the Minoan social system. Trees grew in the valleys, and the hillsides supplied the energy and wood resources for the urbanized communities. Later on in the period, when wood resources diminished, imports from the mainland were secured.

Agriculture, Manufacturing, and Deforestation

On the island of Crete, human transformation of the landscape—such as the cultivation of wheat and pulses plus the rearing of domestic animals—occurred as early as the seventh millennium B.C. at Knossos. Around the fourth millennium B.C., land utilization expanded, and it continued to expand until the second millennium B.C., with the ascendance of Minoan Crete in the eastern Mediterranean (Jarman and Jarman 1968; Van Andel and Runnels 1988). This ascendancy naturally required intensive utilization of natural resources, and the result was the transformation of the landscapes of Crete and of the hinterland that supplied raw materials for its socioeconomic reproduction.

Coupled with the growth in population from the start of the second millennium, along with urbanization and expansion of production and manufacturing operations, the transformation of the landscape, including the consumption of natural resources, was the order of the day. Large volumes of wood were utilized for building construction, manufacturing, and shipbuilding (Chandler 1974, 79). At the height of Minoan power, there was extensive demand for wood to build merchant ships and warships as a consequence of the increased trade between Crete and mainland Mycenaean Greece and the eastern Mediterranean (Meiggs 1982, 97). The size of these ships has been calculated as reaching about 225 tons (Sheratt and Sheratt 1991). Between 1600 and 1400 B.C., the Minoan navy dominated the eastern Mediterranean. Naturally, all these construction activities required natural resources and consequently devastated the landscape.[4] Deforestation occurred to meet the growing wood needs, and when the forests were depleted by the mid-second millennium B.C., wood supplies came from the Greek mainland, such as Mycenae. Over time, the continued reliance on imported wood from Mycenae and Pylos resulted in the transfer of wealth from Crete to the Greek mainland (Perlin 1989, 54).

Surplus generation is reflected in the volume of storage capacity of the palace complexes. In Knossos, the storage capacity for 246,000 liters of olive oil must have required, utilizing modern yields, between 16,000 and 32,000 olive trees to fill the storage jars (*pithoi*). Translated to acreage, this means that approximately 320 hectares were cleared of vegetation to meet this level of production. Furthermore, according to Warren (1985), the large private homes adjacent to the palace complex had their own storage facilities. Estimates indicate that at least 1,000 hectares must have been devoted to the production of this economic surplus. In addition to the production of olive oil, which took only a third of the storage capacity in Knossos, the production of grains such as wheat and barley was also undertaken. South of Knossos, the plains of Messara were devoted to wheat growing. In addition, the plains of Messara were also under the purview of the palace complex of Phaistos. If the entire plain of Messara was utilized for grain production, it is estimated that 2.5 million kilograms of wheat would have been produced per cropping season. However, the linear tablets seem to indicate a production capacity of 775,000 kilograms (Warren 1985).

Agricultural expansion and manufacturing operations also led to deforestation and transformation of the landscape. The introduction of the plow and the differentiation of cultivation areas for crop types further accelerated the deforestation process (Manning 1994). The sum total of this transformation of the land through agriculture, deforestation, and human settlement led to what has been regarded as an inevitable outcome: soil erosion. Soil erosion has various outcomes, among which are the addition of sediment loads to rivers and land deterioration (Van Andel and Demitrack 1990).

The surplus agricultural activity must be placed along with Cretan manufacturing operations within a wider context of the Bronze Age world system. Crete, by the second millennium, enjoyed extensive trading relations with the Greek mainland, the Cycladic islands, Troy, Egypt, Anatolia, and beyond. It was the interactive gateway to the Aegean from the eastern Mediterranean.

At this stage of the second-millennium Bronze Age world system, the extent of the Aegean trading connections can be traced via the tin trade. Because tin is an important component for the manufacture of bronze, the study of the trading of tin provides an eagle's-eye view of the commercial world of the Aegean via its exchange circuit.[5] The circuit extended to Mari, where tin bound for Crete utilized shipping transportation from the Syrian coast (Muhly 1973). Mari, however, was not a producer of tin. It was only a transit stop for the metal. The original source point for the metal was Iran and the Gulf, where it was then transported up the Euphrates to Mari.

Later in the millennium (1600 B.C.), the tin and amber imported to Mycenaean Greece and Crete came from as far away as northern Europe and Cornwall (England). Utilizing the river systems such as the Rhone, trade was conducted between the Aegean and Europe. Greek traders utilizing the river systems met and exchanged products with their Celtic counterparts. The trade consisted of Aegean commodities such as wine, olive oil, and bronze manufactures being exchanged for the raw materials, such as tin and amber, of northern Europe. These resources from northern Europe, including silver from mainland Greece and the Cyclades, were then shipped to Egypt and points in west Asia.

In addition to economic transactions, cultural and social ideas were also exchanged. The duplication of trends and lifestyles by local elites in Minoan society of the trends in the core zone such as Mesopotamia is evident. Cretan palaces by 1700 B.C. bore some architectural similarity, in terms of the monumentality and arrangement of buildings, to that of Mari on the Euphrates. Sherratt and Sherratt (1991, 366) have suggested that the Cretan complexes replicated "many of the characteristics of Levantine ones, both structurally and in terms of common forms of cultural practice and shared elite signifiers: indeed, there is a sense in which the whole of the eastern Mediterranean area, from Crete to the Levant, played a common role in the larger economic system of the early second millennium."

Viewed from this angle, all the palace complexes, including nearby port towns such as Amnisos and Kommos, played key roles in the overall trading system of the eastern Mediterranean and thus of the Bronze Age world system. Bulk imports of copper, gold, silver, lead, tin, obsidian, and ivory were undertaken (Chadwick 1976; Gale and Stos-Gale 1981; Kardulias 1996; Muhly 1973). Ivory was imported from Egypt, while areas of mainland Greece, such as Argolis and the Cycladic islands, supplied metals such as silver and lead (Chadwick 1976; Renfrew 1972; Warren 1985; Weiner 1991). The excavated palace complex of Zakros revealed extensive stores of imported ivory, copper, and semiprecious stones. The manufacturing centers in Crete in turn produced silver and bronze wares, bronze ingot bars ("oxhide" ingots), glass items, pottery, and textiles for exports, besides the olive oil and grains that were also part of the export trade (Sherratt and Sherratt 1991).

Highly controlled textile manufacturing was undertaken in most of the palace complexes (Chadwick 1976). The command and control can be seen in the registering of wool received at Phaistos but recorded at Knossos in the north. Labor for such activity was undertaken by wage workers and slaves; the division of labor separated spinners, carders, and weavers. Furthermore, these workers were also associated by names with particular kinds of textiles (Chadwick 1976). In addition to manufactured products and agricultural produce, wood was also exported from Crete in the early second millennium (Weiner 1991). As indicated by the linear B tablets, poles, masts, and trees were exported to west Asia, Egypt, and Mesopotamia. Various varieties of wood were exported, including willow and elm.

The intensive exploitation of resources for economic transactions impacted on the Cretan landscape. Deforestation generated soil erosion and flash flooding; the latter impacted on the manufacturing processes of Crete. Wood scarcity at Knossos forced changes in production locations or resulted in the closures of facilities. Viewed in this light, severe ecological degradation along with excessive consumption of imported resources such as wood further weakened the political economic strength of Crete by the fifteenth century B.C. Carter and Dale (1974, 68) have even suggested that conditions such as land deterioration could have contributed to the demise of Minoan Crete. By no means were these ecologically devastating trends occurring only in Crete; they were being repeated elsewhere in the system, such as in mainland Greece and elsewhere in Europe. Mainland Greece, which provided wood supplies, was severely deforested, and the areas in northern Europe that supplied minerals suffered scarred landscapes. Such ecological devastation continued well into the Iron Age in Europe, especially in Czechoslovakia, where the degradative impact of the Urnfield settlements on soil formation has been revealed (Kristiansen 1993).

Core–Periphery Relations

From an eagle's-eye view of this part of the Bronze Age world system, Crete during the first half of the second millennium had a full-scale manufacturing and agricultural base participating in the manufacture of value-added products for export. Its command-palace economy was involved not only in the export of surplus agricultural produce such as grains and oil, but also in the export of textiles, metalwork, pottery, woodwork, etc. Such a diversified economic structure thus provided it with a competitive advantage over other regions of the system, such as mainland Greece and the Cyclades.

In this early period, the hinterland supplying Crete was located to its north and west (see figure 3.2). Mainland Greece, the Cyclades, and eastern Europe provided Crete with much-needed metals, such as silver and copper, for its internal consumption and manufacturing needs. A trading network existed from Crete to the islands of the Western String (Schofield 1984). In turn, Crete exported to these areas its manufactured products such as pottery, textiles, grain, olive oil, silver, and copperwares. By the mid-second millennium, following the exhaustion of its wood supply, Mycenaean wood was imported in large quantities by Crete to meet its fuel, manufacturing, and building needs.

Figure 3.2   The Aegean and Western Asia

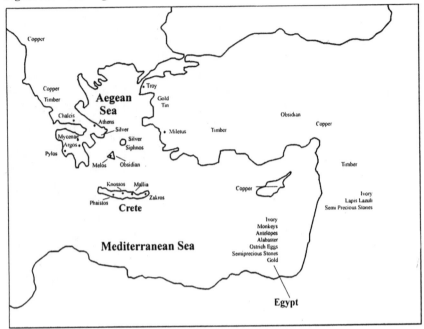

To the east of Crete, its Mediterranean trading partners—which were by this time just as economically transformed as Crete—exchanged for items they could not produce. Items such as perfumed oil, textiles, metalwork, pottery, carved stone, and silver were traded for high-value raw materials such as tin that Crete needed for its manufacturing operations. Across the system, silver coming from mainland Greece and the Cyclades for the manufactured goods and agricultural produce of Crete was then used to pay for the imported metals from west Asia and beyond. By no means was silver the only medium of exchange with these regions. Wood was also exported to Mesopotamia and Egypt.

Based on this extensive network of exchange relationships, some scholars have suggested that Minoan Crete had the status of a thalassocracy (Weiner 1991; Cadogan 1984). Hiller (1984) has even suggested the possibility of a Pax Minoica beyond a thalassocracy. Others, such as Dickinson (1977), have argued against it. Clearly, Crete during the first half of the second millennium B.C. was dominant economically over mainland Greece, the Cyclades, and eastern Europe by nature of the products that were exchanged. It seems that in the Second (New) Palace Period (late sixteenth and early fifteenth century B.C.), there is some evidence of Cretan colonization of the Cyclades and eastern Greece (Branigan 1981, 1984; Cadogan 1984; Manning 1994). Minoan colonization, beginning between 1550 B.C. and 1450 B.C., can be seen in such locations as Ayia Irini, Phylakopi, Kythera, and Akrotiri, located in the Cyclades. Minoan systems of weights and measures have been unearthed, suggesting that these colonies were the gateways for trade to the eastern Mediterranean and Mycenaean Greece, as well as sources for raw materials such as silver. Besides needed metals, horses and slaves were also part of the "commodities" exchanged (Cadogan 1984). Schofield (1982) has further suggested that the trade was organized palatially. On a comparative basis, the trading and social structures established paralleled those we have witnessed in southern Mesopotamia a millennium earlier. With their establishment, socioeconomic lifestyles were reproduced following Minoan lines. No doubt, the exuberant consumption of natural resources followed along with the ecological degradation that ensued.

Such trading/colonization relations underline the connectivity of the hinterland to Crete's manufacturing and production processes. This condition was to change after the mid-second millennium B.C. with the rise of Mycenaean Greece. Unlike the Cycladic colonies, we cannot consider the urbanized centers in Anatolia, Egypt, and Syria, with which Crete enjoyed trade exchanges, as coming under its purview. For these centers were as economically transformed, if not more so, by the start of the second millennium B.C. If one compares the scale and size of these centers with what we have discussed about Crete, clearly Crete's economy and urbanization were by no means on the same scale as Mesopotamia and Egypt. Therefore, its position within the Bronze Age world system by no means is part of the core at this period in time.

CULTURE–NATURE RELATIONS IN MYCENAEAN GREECE

Urbanization and Population

The emergence of Mycenaean Greece as a trading center in the Aegean came almost four centuries later, around 1600 B.C. (Renfrew 1972; Chadwick 1976). This rise placed it in competition with Minoan Crete, which it later came to dominate by the fourteenth century B.C. Mycenaean civilization was centered on the peninsula, with Peloponnese as the major area. The principal places of this socioeconomic region were Mycenae and Tiryns, located in the Argolid, and Pylos, located in Messenia. Other areas included Attica and Boeotia to the north, plus the island of Euboea, and Laconia to the south (see figure 3.3). The climate for this region is typically Mediterranean, with warm wet winters and hot dry summers. Soil compositions in the southern areas are alluvial in nature, especially in the plains of Argos and in Messenia.

Settlement patterns parallel those of Crete and the Cyclades, with the growth factor greater by 3.1 for Messenia, Laconia, and Euboea from the Neolithic era to the early Bronze Age (Chadwick 1972). The growth doubled again in the late Bronze Age, when Messenia and Laconia saw the greatest growth of the three

Figure 3.3   Bronze Age Greece

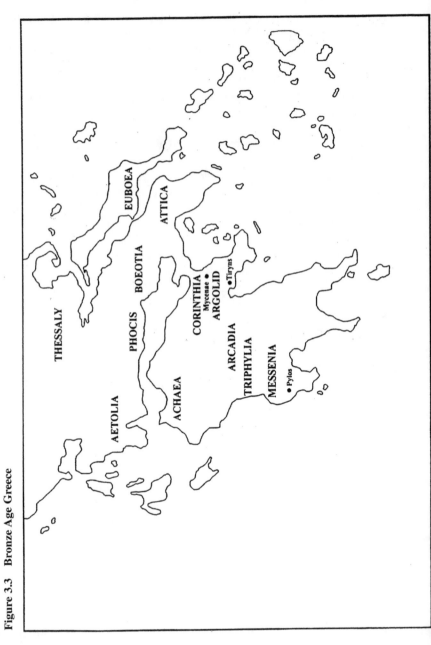

regions. Messenia's growth was exponential throughout the Bronze Age, while Laconia and Euboea had a slowdown in the middle Bronze Age, with growth resuming in the late Bronze Age. Settlement sizes also parallel those in Crete; Pylos occupied about 20,000 square meters, which included the palace and surrounding community. During the late Bronze Age, the largest settlement in mainland Greece was Mycenae, which extended to about 38,500 square meters (Renfrew 1972). When placed on the scale of other urban environments of the Bronze Age system, the sizes of these Mycenaean Greek settlements were rather small in comparison. For example, Mohenjodaro and Harappa, two cities of the Harappan civilization, had areas of 670,000 square meters and 350,000 square meters respectively. In terms of economies of scale based on urbanization, Mycenaean Greece, like Crete, was not the core of the system.

This is also reflected in the population levels, where we find the three regions with a total population of only 272,100 by the late Bronze Age. Messenia's population saw exponential growth, increasing from 23,000 in the early Bronze Age to 178,000 by the late Bronze Age. Laconia and Euboea had a slowdown in growth around the middle Bronze Age and then grew again to 50,000 and 44,100 respectively. Renfrew (1972) has tried to account for this slowdown in the growth pattern of the middle Bronze Age by speculating that coastal settlements were abandoned due to pirate attacks. It is difficult to ascertain whether this was the main causal factor leading to population decreases during the middle Bronze Age. What we need to consider as well is the economic downturn of the Bronze Age world system between 1700 and 1500 or 1400 B.C., when there occurred a slowdown in growth and the demise of urbanized communities and trade throughout the system.

Manufacturing and Trade

Like Minoan Crete, the Mycenaean world saw the building of palaces. The era of palace building reached its height from 1400 to 1200 B.C. (Jones et al. 1986). These palaces were located at Thebes, Athens, Mycenae, Tiryns, Pylos, and Assiros (Chadwick 1976; Jones et al. 1986). As in Minoan Crete, Mycenaean palaces had storage facilities for wine, oil, and grain. Manufacturing facilities were also located within the palace complex. The dominance of Crete in terms of manufacturing and maritime trade in the first half of the second millennium B.C. facilitated the rise of Mycenaean Greece, for Crete drew on its hinterland, of which mainland Greece was a part. Messenia, on the western Pelopennese, benefited from the trade routes going to the western part of the Mediterranean, such as Italy, while the Argolid, to the east, also rose in prominence, benefiting from the trade routes across the Aegean Sea to Troy and onwards to the Black Sea (Dickinson 1977). These routes provided access to the metal-rich areas of the Danube and the Baltic. Notwithstanding the increasing benefits from the trade that passed through mainland Greece, Mycenaean Greece also supplied wood to meet the manufacturing, shipbuilding, and construction needs of Minoan Crete throughout the first half of the second millennium B.C. Mainland Greece, with its bountiful forests, supplied Minoan Crete and demanded a hefty sum for the wood. As a consequence, a transfer of wealth occurred from Crete to the mainland (Perlin 1989, 54).

As a center of accumulation in the Mediterranean, Mycenae by 1350 B.C. had a population comparable to Knossos, and was enjoying an expansionary phase, manufacturing bronze products and pottery. Besides Minoan Crete, Mycenean Greece had trade relations with southern Italy and the Syrian-Palestinian coastal areas (Chandler 1974, 79; Perlin 1989). Its ascendance also led to its drawing on its northern hinterland, such as the lower Rhine. From 1400 B.C. onwards, the Argolid became the core of the Aegean. As in Minoan Crete, the palace economies produced perfume, linen, pottery, ivory, woolen textiles, and bronze products.

In terms of scale of operations, Pylos in Messenia had about four hundred smiths working in the bronze trade (Chadwick 1976). It has also been suggested by Chadwick (1972) that these bronzesmiths were refugees from Minoan Crete after its collapse in the fifteenth century B.C. Linen production at Pylos is estimated at about thirty-seven tons (Chadwick 1976). In addition to these exports, Mycenaean Greece also exported wine, olive oil, and grains such as wheat and barley. The agricultural products and the value-added commodities were exported throughout the eastern part of the Mediterranean, Italy, Sicily, Egypt, and eastern Europe.

Outposts were also established to facilitate commercial exchange and colonial settlement. On a comparative basis, these outposts were similar in function to those established by the Harappans and the Mesopotamians in their hinterland areas. These outposts at Miletus, Trianda, Phylakopi, Enkomi, and Cyprus grew and expanded as a consequence of the increase in trading activities (Vermeule 1960; Immerwahr 1960; Knapp 1993). Like Minoan Crete in the first half of the second millennium, Mycenaean exports were exchanged for imports of much-needed metals such as cooper and tin from northern Europe, the Carpathians, and Etruria (Dickinson 1977, 55, 105).

Such a scale of economic activity and urbanization required intensive appropriation of available natural resources and the intensive utilization of the land. As with Crete, the ecological degradation that ensued, such as deforestation and soil erosion, generated severe pressure to reduce the socioeconomic lifestyle.

Human transformation of the landscape via deforestation and clearing of the woodlands started as early as the seventh millennium B.C. (Van Andel and Runnels 1988). Extensive transformation through deforestation of the landscape occurred later, around 4000 B.C., and continued until the second millennium. The intensity of deforestation in the early Bronze Age caused severe soil erosion on the hills of Berbati and Limnes (Bintliff 1992; Runnels 1995). Over twenty feet of alluvium were deposited on the Argive plain by around 4000 B.C. Intensification of land use and animal husbandry led to severe alteration of the landscape. Population increases along with the introduction of the ox-drawn plow further exacerbated the intensity of land utilization. The Argive plain underwent such transformation beginning about 2000 B.C. onwards (Van Andel, Zangger, and Demitrack 1990).

With the growth of the Mycenaean economy in the second millennium B.C., and the increased demand for woolen textiles, the rearing of sheep became of paramount importance. As in Crete, textile production became a major economic activity in the Argolid. Such growth inevitably affected the landscape. The importation of copper, gold, and other artifacts sparked a need to increase exports such as olive oil and grain (Wells, Runnels, and Zangger 1993). Such attempts to balance import demands generated the need to further intensify agricultural production of cereals and olive oil. Deforestation of oak trees on the hillsides resulted in large amounts of earth and water draining from the slopes onto the plain of Argos and filling up the stream beds, leading to extensive flooding (Runnels 1995). Tiryns and its agricultural lands were affected by flood waters (Runnels 1995). Pylos's harbor was impacted by siltation, as was the island of Melos. With the loss of topsoil, agricultural production in areas such as Messenia significantly diminished.

Following agricultural intensification, terrace walls were built on hillsides to prevent further soil erosion. Other technological solutions were also tried, such as the building of dams to divert water courses and dikes to facilitate drainage (Van Andel et al. 1986; Van Andel, Zangger, and Demitrack 1990). A massive dam was built between the fourteenth and thirteenth centuries B.C., about four kilometers upstream from Tiryns, which was located on the southern edge of the plain of Argos (Kraft, Aschenbrenner, and Rapp 1977; Van Andel, Zangger, and Demitrack 1990). It was one hundred meters in width and about ten meters high (Zangger 1993; Balcer 1974). The effort was made to divert the river, which was transporting tremendous amounts of eroded alluvial deposits from the eastern mountains away from the lower town of Tiryns. In Boeotia, dikes were built to facilitate the draining of Lake Copais. This was also done in the Argolid to contend with winter streams. However, by the late second millennium (1100 B.C.), such efforts began to fail.

Ecological degradation such as deforestation has other consequences. Mycenae, though benefiting from an abundant supply of forested areas, failed to pursue a sustainable yield with its forest-harvesting practices. By the late Bronze Age, when pastureland was cleared in Mycenae for sheep grazing, metallurgical production works had to be relocated to less populated areas where wood supplies were more available. Mycenaean prosperity built on metallurgical and pottery works suffered with the decline in fuel supplies, especially the pottery works at Berbati and Zygories. Population migration followed the closure of these manufacturing centers, and the abandonment of Phylakopi coincided with the deforestation of Melos, where the town was located. The same situation occurred for Berbati, Midea, Prosymna, and Zygories in 1200 B.C., and there were also population losses in

Mycenae, Pylos, and Tiryns. Throughout Mycenaean times, towns and settlements disappeared; for example, in southwest Peloponnese the number dropped from 150 to 14. Other regions experienced similar declines; Laconias, Argolid, Corinthia, Attica, Boeotia, Phocis, and Locris all registered losses (Perlin 1989, 66). Overall, by the eleventh century B.C., the number of inhabitants in Mycenaean Greece fell by 75 percent (Perlin 1989).

THE DEMISE

As we have seen, the pattern of material reproduction of socioeconomic life in the Aegean during the second millennium B.C. followed trends similar to those we observed in the third millennium B.C. in the Mesopotamia-Harappa axis of the Bronze Age system. These trends, such as ceaseless accumulation of capital, core–hinterland relations, urbanization, population growth, deforestation, and intensive land use, ultimately led also to severe constraints on the continued expansion of socioeconomic communities of the Aegean.

Overexploitation of the environment leads to a highly stressed state. Such circumstances as we have seen in the Harappan and Mesopotamian regions coupled with climatic fluctuations and natural disturbances (such as tectonic shifts) could trigger the collapse of the socioeconomic system and lead to structural changes (deurbanization, reorganization of socioeconomic systems, etc.). In the Aegean, the continuous degradation of the environment—starting as early as the third millennium B.C. and persisting to higher intensity in the second millennium B.C. with the rise of Minoan Crete and Mycenaean Greece—laid the structural conditions for the demise of the urbanized communities in this part of the Bronze Age world system. The stressed ecological conditions in combination with socioeconomic forces provide a more comprehensive explanation for the demise of the Aegean communities in the second millennium B.C.

The demise of Crete starting around 1500 B.C. has been widely discussed and debated. Explanations have ranged widely, from tectonic shifts to anthropogenic causes such as the invasion by Mycenaeans, systems complexity,[6] internal rebellions, and revolutionary military innovations (Marinatos 1939; Hooker 1976; Drews 1993; Renfrew 1979; Chadwick 1976; Warren 1985). Single-causal theories of the demise, though elegant and simple in explanation such as that of Marinatos (1939), would need to be scrutinized carefully, as the complex Minoan system could not possibly have collapsed over a single catastrophic event, such as an earthquake on a nearby island close to Crete. Nevertheless, a dismissal of a single-causal factor does not mean that we need to ignore a natural catastrophe as a contributing factor toward the demise. When placed within the context of an already stressed natural environment, natural catastrophes further strained the system. How, then, can we provide some form of explanation for the demise of Minoan Crete and Mycenaean Greece that continues the main theme of this book, that is, ecological limits become also the limits of socioeconomic processes of empires, civilizations, and nation-states, and the interplay between ecological limits and the dynamics of societal systems defines the historical tendencies and expansionary trajectories of the human enterprise.

The Minoans, as discussed above, were intensively and extensively exploiting their island environment as well as its surrounding hinterland. By the beginning of the second millennium, the ecological conditions were stressed severely due to extensive deforestation. Such circumstances, when coupled with natural catastrophes, climatological changes, external invasion, (perhaps) internal rebellions, and the loss of needed natural resources from the hinterland, precipitated the demise of Minoan Crete.

Crete lies on the Aegean plate that straddles the Turkish, African, and Eurasian plates. The island itself is divided by four main north-south faults (Manning 1994). It has been argued by some geologists that between 2800 B.C. and A.D. 400, a particularly active tectonic regime was in force in the southern Aegean (Manning 1994). Such a pattern of geological conditions does provide the grounds for the argument made first by Marinatos (1939) and followed by others, such as Chadwick (1976) and Warren (1985), that volcanic eruptions and earthquakes around 1500 B.C. provided the circumstances for the demise of a complex socioeconomic system. According to Chadwick (1976), the earthquakes followed by volcanic eruption on Thera precipitated further delimiting conditions for the Minoans. The volcanic ash killed vegetation and also destroyed the Minoan naval fleet. The loss of the latter provided the opportunity for the Mycenaeans to invade Crete, since Cretan naval supremacy, which had made Minoan Crete a core in this region of the world system, could then be threatened.

This line of argument needs to be buttressed with exploration of other trends that were occurring within this part of the world system. Climatological changes such as increasing aridity could have caused severe disruptions to food production. Such climatological fluctuations have been proposed as one of the causal factors for the decline of Mycenaean Greece in the late second millennium B.C., discussed later in this chapter. Manning (1994) and Carter and Dale (1974) have suggested that there was evidence of increasing aridity in the southern Aegean circa 2000 B.C. This arid condition—combined with a degraded environment—would have placed severe stress on agricultural production and consequently on the urbanized communities in Minoan Crete, as grain was exchanged for natural resources, needed for the manufacturing processes in the urban environments.

The demise of Minoan Crete around 1500 B.C. should also be considered in the context of the political changes occurring during this time period. The increasing role played by the Hittites and Kassites through their expansion and dominance of Anatolia and Mesopotamia must have impacted on the economic activities of Minoan Crete (Frank 1993). The dominance of the Hittites and Kassites along with the ascendancy of Mycenaean Greece further curtailed the economic reproduction of Minoan Crete. The Minoans, who had relied for their manufacturing processes on the import of necessary natural resources from the various trading interconnections of the Bronze Age world system, experienced the loss of these sources. Those sources located on the Greek mainland came increasingly under the control of the Mycenaeans. Their supply centers in western Anatolia and northern Syria were also restricted because of Hittite expansion and control. Faced with these desperate conditions and coupled with climatological changes and natural catastrophes, the Minoan civilization came under severe stress in its reproductive capacities around the mid-second millennium B.C.

The causal conditions for the collapse of Mycenaean Greece that have been identified repeat what has been proposed for Minoan Crete. Besides the anthropogenic factors such as the invasions by the Dorians and the Sea Peoples and the changing style of warfare in the later period of the second millennium, conditions such as earthquakes, volcanic eruptions, changing climatological patterns, and soil erosion have been advanced as circumstances leading to the demise of Mycenanean Greece around 1200 B.C. (Drews 1993; Betancourt 1976; Carpenter 1968; Chadwick 1976; Desborough 1975). As with Minoan Crete, the natural environment of Mycenaean Greece was by 1200 B.C. severely stretched by agricultural production of wheat, barley, and flax and the rearing of sheep. In addition, deforestation was quite severe; only oak trees remained after most of the pine trees had been cut for internal consumption and export. Betancourt (1976) has asserted that such intensive cultivation was further exacerbated by overspecialization; only one or two types of cereals were grown, which made the economic structure extremely vulnerable to collapse. Placed within such a context, the thesis of climatological change proposed by Rhys Carpenter (1968) for the demise of Mycenaean Greece needs to be considered. Basically, Carpenter's position is that with the shift in the tracks of the cyclonic storms that normally brought rain to Mycenaean Greece, a disruption in the rainfall pattern followed for the interior of the Greek mainland. This resulted in droughtlike conditions during thirteenth and twelfth centuries B.C. The persistent drought was also accompanied by an increase in land temperature. As a consequence, the socioeconomic structure was impacted.

Wright (1968), Chadwick (1976), and Drews (1993), however, have challenged this thesis of climatological change affecting the socioeconomic structure of Mycenaean Greece in the later second millennium. Notwithstanding this, others such as Lamb (1967), Bryson et al. (1974), and Bryson and Padoch (1980) have indicated that the period of drought proposed by Carpenter appears to have prevailed during the time of Mycenaean decline, according to precipitation patterns examined.

For Bryson et al. (1974, 49), during this period (circa thirteenth century), drought conditions and increases in land temperature were also reported in other parts of the world system, further confirming the climatological trend toward increasing aridity. For example, they listed the Hittites' leaving the Anatolian plateau and moving to northern Syria as a consequence of famine. The precipitation was 20 to 40 percent below normal, and temperature was 2.5 to 4 degrees centigrade above normal. Another incident noted during the period was the Libyan people's attempted emigration into Egypt. The precipitation in Libya was 50 percent below normal, and temperature was 1.5 degrees centigrade above normal. Finally, Bryson et al. (1974) also noted that settlements in northern Persia were abandoned due to drought. The precipitation was 50 percent below normal, and temperature was 1.7 to 2.5 degrees centigrade above normal.

Besides climatological changes, earthquakes have also been attributed as causal factors for the decline of Myceneaen Greece. Earthquakes at Tiryns, Mycenae, and the Argolid during the late Helladic Period have been suggested for the decline of these two urbanized communities (Zangger 1993; Mylonas 1966).

Such climatological changes and natural catastrophes as factors leading to demise must be buttressed with socioeconomic variables. By the thirteenth century B.C., Mycenaean Greece was densely populated. The establishment of colonial settlements at sites overseas such as Miletus, Trianda, Phylakopi, and Enkomi must be seen as not only for commerce, but also for excess population to migrate to as a consequence of the inability of mainland Greece to produce enough food to meet consumption. Such inability has its roots in severe soil degradation and in the focus on the production of olive oil, the growing of flax, and the rearing of sheep to meet the needs of the textile industry. As a result, Mycenaean Greece by this time period was not only dependent on imports of essential metals for its manufacturing industry, it was also reliant on food imports. Such reliance would mean that any disruption in the trading routes of the Bronze Age world system of which Mycenaean Greece was a part would mean a severe crisis of socioeconomic reproduction. In the late thirteenth century B.C., there were disruptions in the trade routes as the Hittite empire was collapsing, Troy (Settlement VIIa) was sacked, and Egypt was retreating from Asia (Vermeule 1960). Such transformations would have impacted on Mycenaean Greece, as these areas were suppliers and consumers of Mycenaean imports and exports.

This overall crisis in the system of which Mycenaean Greece was a part resulted in major transformations of the socioeconomic and political landscapes. Depopulation and deurbanization of mainland Greece were the order of the day. What followed was the dark age.

Historians have attributed the onset of this period to several conditions and factors. These have ranged from cultural decadence to invasions and conquests by "barbarians" and nomadic tribes, internal conflicts, overcentralization of authority, famine and disasters, to climatic changes and tectonic shifts (Toynbee 1939; Childe 1942; Snodgrass 1989; Renfrew 1979; O'Connor 1983; Desborough 1972; Carpenter 1966; Bryson et al. 1974; Weiss 1982; Neumann and Parpola 1987; Bintliff 1982; Harding 1982; Schaeffer 1948). Explanations thus have focused overwhelmingly on anthropocentric causes related to social and economic conditions. There seems to be less emphasis on analyzing ecological conditions and the relationship between human communities and the ecological landscapes to account for the onset of the dark age. There are, however, some exceptions. Kristiansen's (1993) analysis of first- and second-millennia B.C. Europe has suggested that ecological relations between human communities and Nature have impacted on their economic reproduction and have also caused socioeconomic organizational changes.

Of the specific dark-age period in question, we find several Culture–Nature trends and patterns exhibiting tendencies that are reversals of the expansionary period: fall in population levels, decline or loss in certain material skills, decay in the cultural aspects of life, fall in living standards and thus wealth, and loss of trading contacts within and without Greece (Snodgrass 1971; Desborough 1972). The archaeological evidence unearthed from this period suggests socioeconomic patterns distinctively different from the style and level of sociocultural life prevailing prior to the onset of the dark age. The pottery and other objects recovered from excavated graves, along with the architecture and design of dwellings from excavated sites, reflect ecological stress and scarcity in natural resources that rendered it impossible to reproduce the earlier sociocultural life during this dark-age period.

Pottery styles became austere, unlike the decadent style of the prior era. The bulk of a pot was usually left plain and in the clay's natural color, and the decorations covered at most a third of the surface area. The lack of intense firing suggests, as well, dwindling energy supplies. As recovery proceeded and the balance of Nature was restored, we find the plain rectilinear or curvilinear patterns in pottery designs giving way to images depicting animals and humans. Such shifts could suggest the return of biodiversity to the environment and reveal to us perhaps the loss of biodiversity at the onset of the dark age.

Beyond pottery styles, other objects recovered indicate a scarcity of natural resources, or that the supply sources have dried up, especially in the area of metals. The use of obsidian and bone for blades and weapons underscores such scarcity and also suggests that trading routes and centers for sourcing the metals might have been disrupted or disappeared. Such conditions do suggest a systemic economic downturn throughout the connective second-millennium Bronze Age world system.

Ecological scarcity required a downscaling of material and cultural lifestyles. Such changes were reflected in burial practices, which exhibited a reorganization of life along modest and rational lines. The designs of clothing and shoes were of the plainest kind. The downscaling process was exhibited further in the formation of decentralized communities and the associated population losses. Whether this lifestyle trend is one that was actively sought as a consequence of ecological scarcity, or whether it occurred as an outcome of the depressive conditions of the dark age is difficult to gauge. What we are sure of is that as recovery proceeded— we begin to witness this by the middle of the tenth century B.C.—trading networks were reestablished and communities were revived. Such an upswing was arrayed with exuberance, materialistic consumption, and accumulation. For during the dark age, materialistic consumption went down, and most of the trading networks disappeared or were restricted only to the area of the Aegean Sea.

What the dark age represented for the region was an era when extreme degradation of the ecological landscape precipitated socioeconomic organizational changes to meet the scarcity in resources in order to reproduce some semblance of the cultural and economic life of prior times. As a consequence, systemic reorganization occurred at various levels, from the way commodities were produced to the clothing fashions and designs. Hierarchical social structures disappeared during the dark age, as made evident by the burial practices, and were restored when recovery proceeded (Whitley 1991). The deurbanization process gave rise to small communities with lower population levels. Seen from the ecological point of view, this downscaling provided the necessary timing for Nature to restore its balance, and for socioeconomic life to start afresh when recovery returned. From these small communities, the preconditions for the rise of the Greek polis was put into play, and what followed was a flourish of political and economic life as soon as the social system recovered (Snodgrass 1971).

Systemic reorganization occurred, but the length of time over which the dark age extended deserves our attention. The fact that it lasted over six hundred years underscores the immensity of the degradation that occurred as well as the extent of time required for expansionary recovery to take place. Seen from a social-evolutionary point of view, a transformation occurred and caused systemic changes that cannot be viewed from a modernist perspective as progressive. What followed in the recovery phase, however, was a sociocultural and political lifestyle conditioned and transformed by the dark age and that formed the basis of Western civilization as we know it today.

In summary, the documentary evidence suggests that during the third and second millennia B.C., the expansionary capacities of the civilizations and kingdoms were connected in a systemic sense to the dynamics of a world system, which by the second millennium was in an ecological depressive long swing (1700 B.C.–700 B.C.). The contemporaneous demise of Harappa and southern Mesopotamia is linked as a consequence of their economic connection in the Bronze Age world system and their comparable ecological relations with the natural environment. The character of these ecological relations is circumscribed by the dynamics of the process of unequal accumulation and the pace of urbanization and population growth that existed then. These three processes (unequal accumulation, urbanization, and population increase) engender ecologically exploitative relations that most often lead to ecological stress. Ultimately, the continued ability to sustain unequal accumulation and urbanization while maintaining a successful balance with the natural environment turned into moments of crisis during the economic downturn of the Bronze Age world system between 1800 or 1750 B.C. and 1600 or 1500 B.C., whereby climatological changes (perhaps even tectonic shifts), ecological stress, civil strife, state rivalry, border incursions, and wars punctuated the historical moment.

During this downturn phase, the peripheral/semiperipheral areas also experienced ecological convulsions. The collapse of Minoan Crete reflects the linkages of this historical world system. The suggested factors for its demise parallel those of Harappa and southern Mesopotamia. Within the ecological depressive long swing (1700 B.C.–700 B.C.), we find, three hundred years later, in another periodic economic downturn of the Bronze Age world economy (1200 B.C.–1000 B.C.), the demise of Mycenaean Greece. Again, the factors for the decline are similar to those that happened in the prior millennium.

NOTES

1. Anatolian influence can be seen in the pottery excavated in the Aegean for this time period (Sheratt and Sheratt 1993).

2. For an extensive discussion of the relations between Europe and the Mediterranean during the Bronze Age, see Sheratt (1992) and Sheratt and Sheratt (1991).

3. Chase-Dunn (1982) has argued that downturns in the world economy, despite the resulting depressive moments for the core, also enable semiperipheral centers to compete and transform themselves.

4. Moody (1997) and Rockham and Moody (1996) have questioned the thesis of the ruined landscape due to deforestation. Their argument is not wholly convincing, as they indicate in their own work the phenomenon of the destruction of the natural vegetation for agricultural cultivation and fuel use in the Bronze Age (Rockham and Moody 1996, 137).

5. Tin was used in the production of bronze, as the resultant alloy is a much superior material for manufacturing purposes. The use of tin bronze started as early as 3000 B.C. in west Asia, with a ratio of nine parts copper to one part tin. This ratio has not changed even in modern-day manufacturing of bronze!

6. Renfrew (1979) has argued that complex early civilizations collapsed when they reached their peak because (1) the costs of available energy and the levels of socioeconomic and political stress became more complex, and at a certain point the benefits outweighed the costs; (2) the territories controlled by the center increased in size, and thus they needed excessive costs to be maintained and they provided breeding grounds for rebellion; and (3) the complexity of the systems made them more rigid to change and thus more liable to stress.

Chapter Four

The Age of City-States: Classical Greece 500 B.C.

Beyond the Neolithic revolution, the process of urbanization has also been depicted as a transformative process in that it initiated a new economic stage in the social evolution of human societies (Childe 1950). As a world historical process, the process of urbanization started as early as the third millennium B.C. in the riverine valleys, such as in Mesopotamia, Egypt, Northwestern India, and China. Over the course of world history, the process of urbanization occurred concomitantly with the accumulation of capital. The distinct socioeconomic structures and composition that marked the urbanization process also entailed a high degree of natural resource consumption to reproduce the lifestyles and socioeconomic practices existing in these cities and towns. For in these cities and towns, social surplus was extracted, accumulated, and consumed. From an ecological point of view, the process of urbanization had significant impact on the environment of the immediate surroundings, and on distant lands that provided the natural resources or manufactured goods to reproduce the urbanized communities.

Besides impacting on the immediate and distant landscapes, the urbanization process also generated the emergence of manufacturing and commerce. The consequence of this is that within the political boundaries of the urban enclaves, pollution emission resulted, which further degraded the neighboring areas and the global environment. Thus, from a world-historical point of view, pollution emissions are as old as the hills.

Paralleling this urbanization process was the growth in human population. Population increases require elevations in resource consumption. Due to this, population is a transforming agent in Culture–Nature relations (Whitmore et al. 1993). With the development of urbanized communities, we also observe an increase in population density. Such increased density required the transportation of products/resources (over vast distances for some scarce products/resources) to reproduce the lifestyles of these cities and towns. In short, besides the accumulation of capital that has engendered severe consequences for the environment, the processes of urbanization and population increases also have borne similar impacts over world history. Urban areas require not only expansion of settlement, but also increases in the cultivable areas to support the consumptive capacities of the urban population.

In the previous chapters, we have focused on the accumulation of capital as one of the main anthropogenic causes of world ecological degradation by analyzing this process in early human civilizations. Concomitant with the accumulation of capital, we have also referenced the urbanization process and population increases as part of the set of anthropogenic causes leading to ecological degradation. For this chapter and the next, our task will focus on these latter two anthropogenic processes of world ecological degradation by examining their occurrences and trends during the classical Greek and Roman periods.

ACCUMULATION OF CAPITAL

From the eighth century B.C. to the sixth century B.C., Greek colonization occurred in the northern Aegean, the Black Sea coasts, Sicily, Italy, and elsewhere in the Mediterranean. The impetus behind this colonization process was primarily for agricultural cultivation and land ownership (Glotz 1926; Hughes 1975). Colonization also increased the volume of trade as the colonies supplied the mother country with cattle and grains, and in return the mother country exported its manufactured products. In addition to the colonial markets, what arose was a Mediterranean market with an "international division of labor" (Glotz 1926, 112). Corn from the colonies in Sicily and Italy was imported to Greece. Libya supplied seasonings, while the Bosphorus sent fish. With regard to metals, copper came from Cyprus, tin from west Asia, and other precious metals from Egypt and Iberia. Luxuries were imported in the form of carpets and embroideries from the East and linens and perfumes from Egypt. In return, the Greeks exported their wines, oils, and manufactured items including pottery. This international division of labor is depicted as follows: "in one place the object was wholesale production of wine or oil, in another foodstuffs were procured in exchange for manufactured articles. While multiplying their relations among themselves, the Greeks reached the markets of the East and of the barbarous countries, from Lydia and Egypt to the Ligurians and the Iberians" (Glotz 1926, 112).

Along the colonization trail west of Greece, the Greeks found well-watered valleys for agricultural cultivation, and the highlands provided grazing land for their herds of cattle and sheep. What we observe is the export of corn, wine, livestock, hides, and wool to Greece, and the import of manufactured goods, pottery, and metals from Greece for colonial consumption plus for the transit trade with the local Etruscans. Towns such as Sybaris and Syracuse grew in size and over time became as large as towns in Greece. Further west were the lands of Liguria and Hesperia, where precious metals were imported by the colonists for transit to

Greece. These activities continued, with the creation of new demands among the locals of these distant lands and the further development of new farms and grazing lands to meet the needs of the cities of the mother country. Forests were cut and mines were developed and exploited to meet the growing trade in the Mediterranean.

To the east of Greece lay the rich agricultural lands of west Asia, along with its rich minerals. Between the sixth and the fourth centuries B.C., widespread maritime trade ensued between mainland Greece and the surrounding areas of Asia Minor with the decline of Crete as a regional power. Large merchant fleets were built and colonies were established for the purposes of trade and "to ensure a flow of essential commodities and materials to the mother city" (Thirgood 1981, 9). Built to guard Greek ships and routes, Athenian war vessels reached an estimated four hundred in number by the fourth century B.C. and eight hundred by the fifth century B.C. Colonization also spread, with settlements around the Black Sea as well as in northern Greece (Meiggs 1982, 121). These settlements became centers of commerce and manufacturing in addition to being agricultural centers; to support these activities they drew on their hinterlands for supplies. All in all, this socioeconomic expansion facilitated the accumulation processes of the city-states of mainland Greece. In terms of timber resource, trade extended to cover the central and eastern Mediterranean including the Black Sea, Asia Minor, and the Caucasus (Thirgood 1981). Greece also exchanged manufactured goods for special wood products from India such as teak and ebony, underscoring the continuity of trading relations with India seen as far back as third-millennium B.C. Mesopotamia. With continued economic expansion, increased urbanization followed, leading to further pressure for increased agricultural production and requiring the colonies in the outlying areas to provide agricultural produce for the urbanized environments. Industrial products were manufactured for exchange utilizing basic primary resources located in the hinterland, thus requiring further deforestation. Hinterland areas such as eastern Greece and western Asia Minor were subsequently depleted of forests. Other coastal regions such as Phoenicia, Syria, Egypt, and Italy provided the grain and food imports for the needs of an urbanized Hellenistic Greece. Of course, deforestation occurred in these areas as well.

This expansionary period generated an increase in the production of manufactured commodities in Greece, requiring basic natural resources such as iron ore for the production processes. The hinterland areas of the Mediterranean such as Asia Minor, Phoenicia, Syria, Sardinia, Elba, and Spain supplied these natural resources. Where iron was exported to mainland Greece, forests were depleted for metal smelting. A million acres of productive woodland were required to meet the needs of a single metallurgical center during the classical age (Thirgood 1981, 56). Furthermore, with developments in the banking system from the sixth century B.C. onwards, profitable investments were made in agricultural production. The rising prices for agricultural produce between the sixth and fourth centuries B.C. and the increasing use of manure coupled with controlled times for plowing and harvesting made farming more profitable and thus facilitated this expansionary process. With production stabilizing in spite of improvements made to agricultural techniques, Greece continued to import grain to meet its food-supply needs in the towns and cities as population density grew there (Garnsey 1988; Sallares 1991). Glotz (1926, 257) has commented that as long as the population of Greece was less dense than that of Attica, grain supply was not an issue. But when the number of consumers who were not producers grew, adequate cereal supply became a problem. For example, Attica produced at most a quarter of its grain requirements and imported the rest. These imports came mostly from Egypt, Sicily, and Italy.

By the middle of the fifth century B.C., with the cities in Ionia acting as its intermediaries, Athens had triumphed over its rivals in the overall Greek trade. Athenian trading arrangements continued along the same patterns as in the previous periods, with yearly volume of imports and exports reaching a total between 30,000 and 32,000 talents (U.S.10,200,000 to U.S.10,800,000 in present-day currency).

URBANIZATION AND POPULATION

The urbanization process is characterized by the formation of the city or polis. The city, a human-engendered structure, has a definitive relationship with the natural environment that surrounds it. This is because to reproduce life within an urban community requires a large amount of natural resources both from the nearby vicinity and transported over great distances. Therefore, cities as human structures reorganize "nature for their own benefit" (Hughes 1998, 106). The growth of cities leads to increases in density of persons living within a given area, thus requiring a tremendous amount of materials and fuels to reproduce life. As exemplified by the modern era, growth in population within an urbanized area also leads to problems of ecological consequences, such as pollution, the need to dispose of waste, and, on occasion, the spread of diseases.

Similar to cities of modern times, ancient cities were the centers of political and economic life and required natural resources to sustain these activities (Finley 1977; Weber 1958). As centers of trade and manufacturing, cities needed natural materials to sustain the activities of production, and also building materials to house these activities. Thus, we observe the long-distance impact of cities upon other areas and regions that supplied these urbanized communities with the necessary natural resources and fuels to reproduce their socioeconomic activities. Ecological degradation was the order of the day in these areas. Within such a framework, we can therefore trace the urbanization process of classical Greece and its impact on the landscape both near and far.

The urbanization process in classical Greece was centered around city-state formation. The polis, or city, was the dominant social, political, and economic institution of the Greeks; it had the power to raise armies, enact laws, conduct treaties, and collect custom duties. By the classical period, there were about 750 of these polities. The largest of these cities was Athens, which by 430 B.C. had about 100,000 to 155,000 persons living within its city limits, and about 160,500 persons living in the 2,600 kilometers of land surrounding it (Sallares 1991; Gomme 1933).

Planning and building within the Greek polis followed a distinct pattern, which also became the blueprint for cities in the colonies of Asia Minor, Sicily, and Italy. The nucleus of the city was the acropolis, a hill that was defensible and, in earlier times, the seat of the king. Outward from this center were clustered the rest of the buildings. A pattern of cities evolved in the shape of a wheel, with the acropolis as its hub. In the lower part of the city was the agora, which included the marketplace, as well as the sites for social, political, and religious life (Wycherley 1967). As the city grew, the agora contained bigger and better buildings. Mature cities had a council house, called the *prytaneion*, where individual magistrates had their offices and also where records were kept. The increasing complexity of business led to the construction of buildings for bankers, merchants, and speculators. Here were located the storehouses for goods that were imported or exported. In the marketplace stood rows of shops for the sale of manufactured products and other items of consumption. Also included in the agora were altars and temples for religious practices. Statues of gods and heroes dotted the area, along with columnar facades fronting the buildings. Such an impressive array of structures naturally required a tremendous amount of natural resources to build and maintain it. The streets were regularly spaced and straight, and they crossed each other at right angles. The main thoroughfares were lined with the houses of shipowners, businessmen, and bankers. The side streets were the homes of those on the lower end of the economic ladder.

Wood, limestone, and marble were the materials most commonly used for construction of these buildings. Copper rivets were employed to anchor walls and ceilings (Coulton 1988). As the cities developed, the public buildings became grander and grander, and more marble was used in their construction. Mounts Pentelicus and Boetia provided the marble. Timber came from Attica, Crete, Egypt, and Euboea, and in certain cases, from as far away as northwestern India. Mud bricks were mostly used for the building of homes, which rested on a low foundation of stone to prevent the moisture from seeping upwards. The level of utilization of resources to fashion these buildings was exorbitant. For example, a single kiln for the production of lime required for building consumed a thousand mule loads of juniper wood for one burn; six thousand metric tons of wood were needed yearly to supply fifty kilns (Wertime 1983).

As centers of commerce, Greek cities housed the various industries that formed the backbone of exports to the colonies and the world markets. Manufactured items such as swords, woolens, leather goods, and pottery were produced in cities located in Attica, Boetia, Thessaly, Phocis, Corinthia, Laconia, Argolid, and Messenia. The production of such items, especially metallic products, required not only the raw materials that would be fashioned, but also a substantial source of energy. Most of the energy was derived from wood. This had considerable impact on the ecology of the landscape. We witness this in the case of Athens.

By the beginning of the fifth century B.C., mainland Greece was on a socioeconomic growth trajectory, especially the city-state of Athens. Becoming a center of accumulation also required a strong navy to ensure control of the trading route and to thwart aggression from Persia. Because the Persians controlled much of northern Greece, Athens initially had to rely on local timber for its shipbuilding program. By 357 B.C., it had an inventory of 285 triremes (Meiggs 1982, 123). With the defeat of the Persians in 469 B.C., Athens' position as a center of accumulation in the Mediterranean was ensured, and socioeconomic expansion ensued. This resulted in rapid urbanization, which required large quantities of wood for buildings and houses. By this time the city's population, including surrounding areas, had grown to nearly 200,000 (Chandler 1974, 79). This demographic surge, as in the previous periods in world history, further increased the demand for wood as fuel (charcoal) and for the manufacture of commodities (Perlin 1989, 86). As a consequence, the price of wood rose. This prompted the Athenians to search for other wood sources through conquest and colonization. One such source was Amphipolis; in 495 B.C. Athens colonized the city and began exploiting its forests. The extensive use of timber for shipbuilding to fight the Peloponnesian War had led Athens to cut down its surrounding forests and to rely on Amphipolis (Meiggs 1982; Perlin 1989). When this source was cut off, Athens relied on Macedonia (Meiggs 1982).

The above impact was furthered with the population increases that characterized the booming economic expansion during the fifth and fourth centuries B.C. Slaves and metics (noncitizens) staffed the production lines. The Laurion silver mines that supplied Athens with silver required at least eleven thousand workers. The total population of citizens in Greece during the fourth century B.C. was around two million (Sallares 1991). We witness the population levels at their highest when Greece was at its height of power during the fifth and fourth centuries B.C.; the population levels declined in the Hellenistic period that followed. Population levels of Attica showed this trend of decline. In 431 B.C., Attica had a population of 315,500 persons; it dropped to 218,000 by 425 B.C. (Gomme 1933, 26). If we count only citizens, the drop was even greater: from 172,000 in 431 B.C. to 84,000 by 313 B.C. The prior century of exuberant growth must have led to ecological landscape degradation, which over time bore consequences on population levels. For example, the decline of availability of wood for fuel led to the relocation of metal industries, which meant the loss of jobs. Such impacts will be discussed in the following section.

ECOLOGICAL DEGRADATION AND POLLUTION

The extent of the degradation of the landscape and the atmosphere during the classical Greek period was extensive. Urbanization, population increases, and the process of accumulation of capital were the major reasons for the degraded landscapes. Not only did the degradation bear impact within Greece, but it extended to areas where the Greeks had colonies and trading relationships. It also led to pollution of the cities' air and rivers as a consequence of the manufacturing processes and the density of human population. The impact thus was also extensively spatial in character, as the economies of the Mediterranean, Europe, and west Asia were linked horizontally in the overall process of the accumulation of capital on a world historical basis.

Deforestation

One of the most common forms of degradation of a landscape is deforestation and its consequent aftermaths such as soil erosion and the siltation of rivers. Deforestation in the classical Greece period had several origins. Both within Greece and within the Mediterranean and west Asia, deforestation resulted from the clearing of fields for agriculture, obtaining wood for fuel to meet human consumptive and manufacturing needs, and producing timber for construction of buildings, homes, and ships. Military campaigns also generated deforestation in areas where major land wars were fought or where armies marched.

To meet the growing cereal needs of the urbanized communities of classical Greece, farms were established primarily through colonization not only on mainland Greece but also to the east, in Asia Minor, and to the west, in Sicily and Italy. Forests were cleared for agricultural lands. Lowlands lost their trees before the highlands, and forests that stood closer to rivers were exploited before those that were farther away. Overgrazing of deforested land by cattle, swine, and sheep ensured that deforestation became permanent as the urbanized communities required not only cereals but also meat. Sheep also provided the wool for the textile industry, located in the urbanized communities of Greece. Furthermore, animals such as pigs ate acorns, chestnuts, etc., thus further destroying the ability for regeneration.

Deforestation can force the relocation of industries, which over time may lead to the loss of a region's commercial and productive dominance. By the fourth century B.C., Athens experienced this fate when it faced shortages of wood and thus had to relocate its metal industries closer to sources of wood. Denied further access to other forestland in Greece, Athens continued to face shortages, and its power on the Greek mainland slowly declined (Perlin 1989, 100).

Soil erosion, an aftermath of deforestation, generated severe consequences for reforestation. The torrential rains washed away the soil and thus inhibited the regeneration of the forests in the uplands of Greece. Silt, sand, and gravel then were deposited into the rivers and estuaries. Devoid of trees, these areas also suffered severe flooding, as trees and plants and their root systems normally held water, releasing it year-round into streams and rivers. Soil erosion transformed the coastlines of Greece, and, in places, dams and dykes had to be built to channel the silt and flooding. The cities of Greek Asia Minor, such as Priene, Myus, and Ephesus (fifth century B.C. to second century B.C.), over time became inland cities, where once they had been located on the coastline of Asia Minor facing the Aegean Sea.

Deforestation also caused what are now known as microclimate changes in the specific areas where severe deforestation occurred. Temperature increases at the local level were recorded. Theophrastus recorded "changes in local climates that he himself observed: after the trees had been cut down around Philippi, for example, the waters dried up and the weather became warmer" (Hughes 1994, 85).

Diseases such as malaria became more abundant. The anopheles mosquito could breed in the new marshes that were created by runoffs from deforestation. These diseases impacted on the population and sometimes caused population loss. Since ecological degradation tends to peak at the height of economic expansion, the ensuing phase of an economic downturn would not only usher in economic distress, but also witness a higher rate of diseases among the population. Thus, indirectly, ecological degradation could produce conditions for diseases to emerge.

Mining, Pollution, and Species Extinction

Besides deforestation, mining for natural resources to meet the demands of manufacturing and commercial life in the urbanized communities also engendered landscape destruction. Gold, iron ore, and silver were actively sought. With the widespread use of coinage from sixth century B.C. onwards, silver was in high demand. Silver was mined at Laurium, gold at Mount Pangaeus in Macedonia, iron ore in southern Laconia, tin in England, and lead, asbestos, and asphalt in various parts of Greece. At Laurium, more than two thousand mine shafts were dug, some to the depth of 120 meters (Freeman 1996, 171). Mining activities involving pits and tunnels further exposed the land to soil erosion. The tunnels at Laurium were more than eighty-seven miles long (Hughes 1994, 118). Mining also contributed to deforestation. Furthermore, in certain cases, mining led to the leaching of chemicals from the soils into the water more rapidly than normal. This resulted in the pollution of rivers and groundwater by lead, mercury, and arsenic, which were generated by hydraulic mining or were leached from the mines.

Pollution was not limited to the rivers and ground water. Air pollution was also a major outcome of urbanization and manufacturing activities. During the classical period, urban life required the consumption of energy, and the utilization of wood in the form of charcoal was the order of the day. Smoke, dust, and odors emitted from charcoal braziers and oil lamps. The manufacturing sector also emitted its share of pollutants. The wide-scale use of wood as fuel generated air pollution. Limestone kilns required a tremendous amount of wood for their production of plaster and mortar, necessities for building construction. With all this pollution, air quality was poor in the urbanized areas, especially during temperature inversions that held the smoke and dust in suspension.

Similar to what is experienced in the modern day, species extinction was an issue in the classical Greek period. By the end of the Hellenistic period, the lion and the leopard were eliminated from Greece and western Asia Minor. Other predators, such as wolves, lynxes, and jackals, were made scarcer; so were other birds and animals. Hughes (1975, 1994) has suggested that these losses were a consequence of the spread of agriculture, grazing, deforestation, and hunting. Naturally, urbanization and population increases were vectors that spurred further agricultural production, grazing, etc., which in turn led to diminishing species on the Greek mainland and in other areas of colonization in the Mediterranean.

Urbanization in classical Greece transformed the landscape in other ways. Cities needed water, which was taken from nearby rivers, streams, and wells; when more water was required aqueducts were constructed to tap the supplies of the countryside. Water was channeled from Mount Pentelicus to Athens and Piraeus. Such channeling drained the water sources of the countryside, which generated consequences for the region's biota. The impacts of these practices were especially felt in the dry spells that Greece annually encountered.

Conservation Measures

One of the major threats facing classical Greece was deforestation. With increasing wood shortage, conservation measures were introduced, such as solar heating (in the form of passive solar design of houses) and the utilization of fuel-efficient braziers. The use of solar energy to heat homes was a response to the energy shortage (Butti and Perlin 1980). Excavations show that architecturally designed solar homes flourished in Greece. Passive solar design involved an open court built at a structure's southern end, with a room on the northern end of this court. The southern orientation allowed for heat from the low winter sun to dissipate to the room located at the north end of the courtyard. Cities such as Pirene and Olynthus were designed to take advantage of solar heating.

Such initiatives were not the only ones tried. Besides attempts at the replanting of trees by private farms, the state also introduced a series of measures through the establishment of regulations for timber harvests and the construction of public works to control drainage and erosion. Besides the state's initiatives to prevent further soil erosion, private farmers also practiced contour plowing and terracing in their agricultural practices. Shoots of the first crop were plowed into the ground to increase fertility. Stone walls were erected to prevent soil erosion. Regulations were enacted to protect sacred groves, indicating a pious appreciation of trees and to ensure that they were not cut. These regulations even prohibited the removal of dead wood; penalties were imposed for violating them (Hughes and Thirgood 1982a). In certain cases, tree plantations were encouraged in Egypt. Plantations were protected by laws, and animals were prohibited in them.

In view of the facts laid out in the preceding pages, Childe's (1950) thesis of the urbanization process as a transformative one that initiates a new economic stage in the evolution of human communities needs to be considered in the context of the negative effects this process has on the ecological landscape. Urbanization as a transformative process in classical Greece was very ecologically degradative. When combined with the concentration of dense population involved in the accumulation of capital, the urbanizing process intensified the destruction of the landscape and also cast ecological shadows in outlying areas that had to supply the natural resources and energy to reproduce the urban communities.

As we have mentioned in earlier chapters, this urbanizing process by no means is limited only to the Mediterranean communities. It is a trend that started much earlier and continues to the present day. In the subsequent chapter, we will follow this transhistorical process to Rome. With the rise of Rome and the reach of its imperial rule, the urbanizing trend was extended and intensified throughout the Mediterranean, North Africa, Europe, and the British Isles.

Chapter Five

The Age of Empire: Rome A.D. 200–A.D. 500

From Mesopotamia and Harappa through classical Greece, the urbanization process continued unabated until the era when Rome was supreme in the Mediterranean world. This urbanization process proceeded further with Rome's rise as an imperial power in the Mediterranean, and with its extensive trading connections with areas in northern Europe, the Mediterranean, Africa, and the Far East that were not under Roman territorial rule. In the Far East, Roman trade reached as far as China and Southeast Asia. With rapid urbanization and economic transformation, population increases also followed. Thus, what transpired in classical Greece was repeated in Rome from the first century A.D. onwards.

Economic transformations and exuberant lifestyles led to increased consumption of luxuries and commodities that necessitated the importation of natural resources and consumer items from different parts of the world economy. Naturally, the landscape and other living beings were impacted by such trends and tendencies. Therefore, besides Rome's immediate surroundings in Italy, degradation of Nature continued in other parts of the world that had trading relations with Rome. Roman imperial expansion via wars also caused severe degradation of the environment in the form of deforestation (Perlin 1989; Hughes 1994). In the case of classical Greece, deforestation—such as in Amphipolis—occurred mostly as a consequence of Athens' building naval ships for its wars and to achieve naval supremacy. In the case of the Roman military, besides cutting down trees to obtain wood for shipbuilding, Roman legions chopped down forested areas where they camped. The deforestation ensured that the forests could not provide cover and camouflage for Rome's enemies (Perlin 1989). Areas of considerable size were deforested; an army of three legions would require a defensive rampart enclosure of about 2,000 by 3,000 feet when it halted its march and camped (Dornborg 1992).

ACCUMULATION OF CAPITAL AND TRADING RELATIONS

By the closing decades of the first century B.C., the Mediterranean and its surrounding littoral were a Roman world. Trade routes crisscrossed the region in every direction, and the major ones led back to the city of Rome. These routes spanned eastward across Asia Minor to China, and southeastward via the Gulf to India and Southeast Asia. Rome by this time was highly dependent on imports of not only natural resources, spices, animals, and other exotic products, but on grain, such as wheat, from its colonies and distant lands. The volume and velocity of economic activity, and thus the accumulation of wealth, was heightened as the Roman world expanded in terms of political control and cultural exchanges. This control lasted for more than five hundred years. A large part of continental Europe came under Roman control, and northwestern Britain was also added. To the northeast, territorial gains were supplemented by the addition of regions such as Dacia. Besides territorial control, the Roman state also claimed for itself the natural resources, such as the mineral resources and stone quarries, of the conquered territories. As early as the second century B.C., the overseas provinces of Rome provided significant revenues in the form of tribute to Rome based mostly on levies imposed on land, population, mines, and transport and harbor dues. Thus, perennial revenues were generated to provide for the needs of the Roman state and its elites (Nash 1987; Hopkins 1980). Such levying of taxes accelerated further commerce and economic exchange of manufactured items in the provinces through increases in agricultural production to meet tax payments (Hopkins 1980). Economic activity and the accumulation of wealth occurred in several ways. Taxation or levies were sometimes collected through direct acquisition of agricultural produce and materials. This fashion of tax collection was often established to provide for the requisition needs of the provincial Roman administrators and of the Roman legions stationed along the borders. Where the taxes were collected in cash, commerce expanded as a consequence of market exchange that resulted to generate the payment in silver coinage. Thus, an increase in agricultural production occurred, and in the urban centers, manufacturing expanded as a consequence of tax collection. Trading networks emerged all over the provincial territories, even penetrating into areas that were beyond Roman territorial control. The networks spurred the expansion of manufacturing and extractive industries related to leather, wool, pottery, and other natural resources, such as scarce metals, that Rome lacked (Haselgrove 1980; Hedeager 1980). In all cases, the beneficiaries of the taxation schemes and expanded economic activity were the elites in the provincial urban centers and, of course, in Rome itself.

Trade routes underline the connections between the Roman world and other distant lands that were not under Rome. The type of product exchanged within the Roman world, and also among the Roman world and other distant lands such as India, China, and Southeast Asia, emphasized a division of labor between more technologically developed areas and those that were less so. It also suggests an economic system that was connected across regions and that saw trade imbalances (deficits) during certain periods. Such integration has led to suggestions that the ancient economy was also subjected to economic long waves of at least the Kondratieff kind. C. J. Going (1992), using pottery production as a proxy for rise and decline of economic expansion and contraction, has argued along such lines.

The trading routes extended across land and sea, and a careful delineation of these trading relations underscores the high division of labor that occurred in the manufacture of items not only for exports, but also for exchange. Geographically, from Rome, the trade routes to China, India, and Southeast Asia branched off into three circuits. The northernmost circuit ran via the Black Sea through Byzantium and central Asia. The central route went via Syria through Antioch and the Euphrates to the Persian Gulf and beyond. The southern circuit was through Alexandria, northern Africa, and Petra via the Red Sea and the Nile and beyond. For its trading relationships with Europe and the western part of the Mediterranean, Rome's trading circuits spanned to the West via the sea to Spain, and to the north via the Rhone-Saone rivers and waterways to Gaul, northern Europe, and Britain.

The complexity of these trading routes is distinguished further by circuits that radiated from these main routes at the local and regional levels (see figures 5.1–5.3). For example, the Red Sea and Arabian Sea region underscores such local complexities of trading exchanges and entrepôts (see figures 5.1 and 5.2). The complexity is defined by the different trading items that were

Figure 5.1 Trade Routes: India, Persia, and Arabia

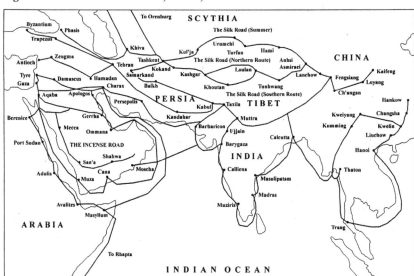

Figure 5.2 Trade Routes: East Africa and India

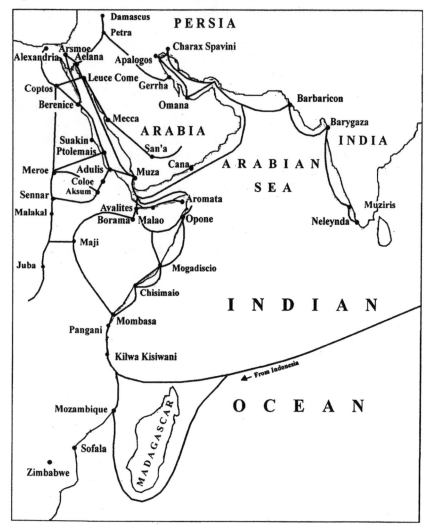

exchanged and transported across the trade circuits depicting the world economy at this point in time.

The Eastern and Western Trade Routes

The easternmost point of the trade routes radiating from the West was Loyang, China. This route, the Silk Road, in the first century A.D. stretched from China—traversing several political entities such as the Kushans and the Parthians—to the

Figure 5.3   Trade Routes: China and Southeast Asia

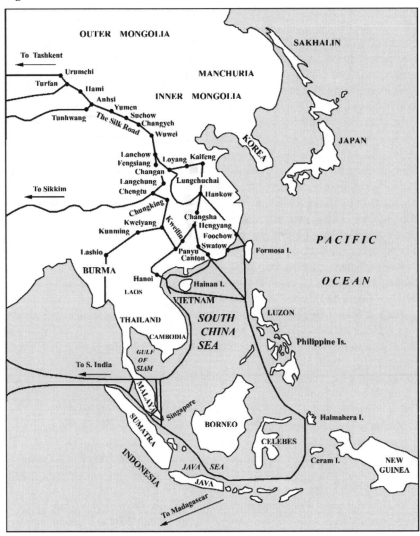

West and ended at Antioch via Seleuceia and Ctesiphon. Along this 4,950-mile road were transported products from China such as silk, ginger, cinnamon leaves, cassia bark, and other spices and semimanufactured goods such as steel and ink (Miller 1969). From the West flowed gold and silver in bullion and coinage, glass, fine cloth, and other fine manufactured items.

The central and southern routes primarily connected the West with India, the Arabian Peninsula, Egypt, and Africa. The southern routes connected Rome to Egypt, Africa, India, Sri Lanka, and Southeast Asia. The port of Alexandria was the connective point and acted as collection center (entrepôt) for products from the East. Other ports on the Red Sea coast were also very active in this overall trade. From the late first century B.C. to the second century A.D., the ports of Clysma, Myos Hormos, Philoteras, Leukos Limen, Nechesia, and Berenice on the Red Sea were on the trade routes (Sidebotham 1991). From China, India, Sri Lanka, and Southeast Asia, steel, iron, fine muslin, silk, pearls, teakwood, ebony, ivory, tortoise shell, cinnamon, cloves, nard, cassia, exotic animals, precious stones, cotton, millet, frankincense, myrrh, and even slaves passed through these ports on the way to the Mediterranean (Miller 1969; Deo 1991; Stern 1991; Carswell 1991). Iron was mined from early Han times in the provinces of Shensi, Hupei, Kansu, and Szechwan. Animals were also imported to the West to meet the exuberant needs of the Romans. They ranged from parrots and cockatoos to tigers. Several ports on the west coast of India emerged to become part of this trading circuit. They included Barbaricon, Barygaza, Muziris, Nelcynda, and Becare. Mantai in Sri Lanka was also a major conduit of products from Sri Lanka to the Persian Gulf and beyond. In return, Rome sent gold and silver bullion, plate, and coinage, glass, barley, wheat, sesame oil, wine, fine silk and cotton cloth, bronze statuettes, cameos, mirrors, and pottery to pay for these imports. India was primarily a source of natural resources and manufactured products, whereas Southeast Asia and Ceylon provided wood, spices, and products from the sea.

The southern trade route, which covered Africa, had Alexandria as its main entrepôt. Grain, wood, oils, exotic animals, and slaves were the main exports from this region to Rome. From the first century A.D. onwards, Rome imported at least fifteen million bushels of grain per year from North Africa and Egypt (Casson 1954). By the fourth century A.D., Africa became the principal source of grain and oil for Rome (Fulford 1992).

The western trading circuits of the Mediterranean and continental Europe covered mostly the exchange of natural resources and raw materials. Of major importance were the gold and silver that Rome needed to balance its trade deficits with the East. The gold, silver, tin, and other metals were sought after in the mines of Spain, Gaul, and Great Britain (Nash 1987). Wine and oil from Spain, France, and Greece were also sent to Rome (Blazquez 1992).

The overall trade was conducted via land and sea. Over land, various types of draft animals were utilized, such as the Bactrian camel. By sea, merchant shipping was employed. The utilization of the sea required an intensive consumption of wood to build ships. By far, using the sea was an economical way of transporting bulky and heavy items such as grain, oils, and wine. The cost differential was high. For example, the difference was about 60 times more for transporting an item by land than by sea from Syria to Spain (Hopkins 1978). On the major runs, such as the grain route between north Africa and Rome, ships of considerable size—measuring two hundred feet long and capable of transporting twelve hundred tons—were used. Some of these ships carried passengers; it is estimated that they could carry up to six hundred persons (Casson 1954).

Roman Exports

Roman imperial expansion provided Italian producers with trading opportunities (see figure 5.4). There was a demand for Roman goods created by the Romanization of the conquered territories. The communication infrastructure established by Rome and the presence of Roman legions in the occupied territories facilitated the consumption of manufactured goods (Whittaker 1994; Oldenstein 1985). If one considers Rome the core of the world economy, its manufactured items such as pottery, glassware, perfumes, red coral, jewelry and cut gems, and textiles (such as chrysotile) were items sought after in other parts of the world and in the conquered territories and colonies. Notwithstanding this, the production process was not as centralized as we might envisage; local manufactured goods such as pottery and other items produced in the outlying areas of the Roman empire were also sources of supply for local consumption (Nash 1987; Hopkins 1980).

Besides the locally produced manufactured items, other goods were exported to Spain, Gaul, Europe, India, Southeast Asia, and China. Where the local industries were equally advanced, such as in India and China, Rome exported red coral and fine textiles, which were in high demand, or had to pay for its purchases using silver and gold bullion or coinage. The precious metals of silver and gold were extracted from Rome's western conquered territories such as Spain and Gaul. In turn, Rome sent its manufactured items (Thorley 1971). Increasing consumption from the first century A.D. to the third century A.D. necessitated an increased in imports

Figure 5.4   The Roman Empire: Principal Supply Routes to the Northern Frontier

of spices, exotics, and other items from the East that were paid for with silver and gold. As a result, by the first century A.D., Rome experienced a trade deficit with the East. Annual loss to India alone amounted to fifty-five million sesterces. It is estimated that annual trade deficits with India, China, and the Arabian peninsula amounted to 100 million sesterces combined (Miller 1969; Hopkins 1988). This is equivalent to about twenty-two thousand Roman pounds of gold.

To meet the growing needs of silver and gold as payments for imports from the East, and also of iron and manufactured items for sale to the rest of the conquered territories and colonies, mines were developed in Spain, Gaul, and Great Britain. Spanish mines in Galicia, Lusitania, and Asturias produced about 10 tons of gold each year (Howgego 1992). Total consumption of silver for coinage in one year by the Roman mint was about 40 to 45 metric tons in the first century A.D. The mines ran as deep as 250 meters underground. According to estimates, each metric ton of silver required five hundred to a thousand work-years, and each ton of pure silver involved the removal of a hundred thousand tons of rock. Tremendous amounts of oil were needed to light the underground mines, and ten thousand metric tons of trees were needed to smelt the ore to produce a ton of silver. The byproducts of such smelting yielded about 400 tons of lead and slag that had to be disposed of. More than forty thousand slaves were used to work the silver mines in Carthago Nova in Spain. Roman iron mines in southern England produced about 550 tons from six sites during second-century A.D. (Hopkins 1988).

Manufactured items were produced mainly in the urban centers of Rome and in the outlying territorial boundaries and colonies such as those in northern Italy and Gaul. The quantity of items produced for consumption and export was quite extensive. In the first century A.D., for example, in southwestern France, pottery production occurred on a considerable scale. Hopkins (1988, 763) notes a production schedule that listed more than seven hundred thousand items. The manufacturing center at Rheinzabern in southwest Germany produced a million pieces a year (Garbsch 1982). Naturally, such scale of production required the consumption of natural resources and energy. In the case of pottery and metallic works, wood was the main form of energy in the production process. Other manufactures, such as fine textiles, that were sought after in other parts of the world economy required the importation of silk, cotton, wool, and flax as raw materials. Roman jewelry made from imported gemstones and red coral were luxuries in demand in other parts of the empire and also in the East.

URBANIZATION

As Moses Finley (1977, 305) has surveyed, the Roman world was a "world of cities." Notwithstanding the importance of the agrarian sector of the Roman economy with the villa as a focus of rural production, the population in the countryside also lived in urban environments of villages and towns instead of in isolated farm housing. Therefore, the Roman urban environment was very similar to the world of classical Greece, at least in terms of urbanizing trends. From republic to empire, urbanization continued uninterrupted, spurred on in the latter period of "High Empire" (first two centuries A.D.) by the increase in the volume of trade. The level of urbanization was not to be surpassed until a millennium later (Hopkins 1978).

Besides the growth of towns and cities as a consequence of the growth of trade and manufacturing, urbanization was also an outcome of Roman imperial expansion and colonization. We witness the rise of large towns and cities throughout the Roman empire—especially in the outlying border regions—not only because of tax collection and trade expansion, but also toward ensuring imperium. The latter required the fortification of necessary defenses to keep the territorial boundaries of the empire intact through the building of forts and the establishment of associated towns.

The imposition of tax collection on territories under Roman control generated the stimulus for further urbanization, in addition to the urbanization trend that was spurred on by economic expansion. The former stimulus, an outcome of imperial expansion and colonization, engendered an increase in manufacturing and trade based on the way the Roman world was structured. Between 200 B.C. and A.D. 400 the Roman world could be divided into three spheres (Hopkins 1980). Basically, the inner sphere comprised Rome and Italy, where the surplus generated throughout the empire was mostly sent. Surrounding this inner sphere was a ring of resource-rich provinces such as Spain, Syria, Greece, Gaul, North Africa, and west Asia. This group was a net exporter of surplus, and its exports were consumed in the inner sphere. Finally, the third sphere was composed of a ring of frontier provinces where the Roman legions were stationed for territorial protection and enforcement of imperium. The second sphere—the ring of resource-rich provinces—paid their taxes in money form (silver coinage); in order to generate the income for payment, agricultural produce (grain, wine, oil), natural resources, wool, hides, and other manufactured items (cloth, dyes, ropes, etc.) had to be produced for commercial exchange. The frontier provinces on the whole tended to pay their tax revenues in kind, i.e., in the form of agricultural produce or goods. Thus, across the two outer spheres, expansion of economic activity and trade occurred in these manners. All these types of economic activities proved quite draining on the ecological landscape.

Such a growth of economic activity and trade occurred within the confines of towns and cities throughout the Roman empire. Urban centers emerged as collection points for staples such as grain, wine, oil, wool, flax, hides, and metals. Others also emerged to house the manufacturing of cloth, leather, dyes, ropes, pottery, etc. Provincial centers such as Narbo and Tolosa in Gaul and Aquilea on the Adriatic became large commercial places, and they even served trade beyond the boundaries of the Empire (Nash 1987). Urbanization along these lines was fostered further by the emergence of intermediate markets and by the increase in the number of Roman officials and administrators, in addition to the Roman soldiers stationed in the provinces. Associated with this increase in people involved in governance and control, artisans and other workers also swelled, primarily to

serve the needs of the former. In this light, urbanization continued on its historical growth trajectory. The Roman world thus followed the spectrum of "urban
revolution" that has emerged in various places throughout world history since the
dawn of time—a feature typical of human civilization following the advent of
surplus generation (Childe 1950).

Cities, Towns, and Villas

The apex of the urbanization process was Rome. By the time of the late republic,
no city in the Mediterranean or in Europe would surpass the city in size and scale
of consumption (Nash 1987). By the first century A.D., it had an estimated population of about one million (Hopkins 1978), the size of London eighteen hundred
years later. Paralleling this scale to a lesser magnitude were cities in the provinces
and territories such as Alexandria, Antioch, and Carthage, with populations
around three hundred thousand and comparable to other European cities almost
two thousand years later.

Rome as the center of administration became the main locale for both private
and public expenditures. It drew materials, labor, and foodstuffs from across the
Roman world and housed a wide range of manufacturing and export industries.
As a result, its needs were extensive and expansive. These ranged from food for
its population to building materials to support its manufacturing and military enterprises (Duncan-Jones 1990). This is clearly an exemplar of how an urban community through its needs transformed its immediate ecological surroundings and
other distant landscapes from which it drew its resources.

As the capital city, Rome was embellished with grandiose public buildings and
clustered with multistoried structures housing places of business and manufacturing. It was not only a commercial city; it was also a place where manufacturing concerns were located. The public buildings were clustered around the forum
with its public assembly area, the comitium, and nearby lay the curia, the meeting place for the Roman senate. Within this complex of public structures were
also located the various temples and fountains. These buildings and others exceeded in size any other buildings constructed before the High Middle Ages
(Duncan-Jones 1974). Besides the size, the materials utilized and the ornaments
affixed to these structures were extravagant (Duncan-Jones 1974; Packer 1988).
Construction costs ranged from as high as two million sestertia to about 100,000
sestertia around A.D. 200 (Duncan-Jones 1974). Costs in the provinces as re-
trieved from surviving records was much lower. The median cost was about
43,500 sestertia. These lower amounts do not automatically suggest that construction costs were lower in the provinces. The paucity of sources for building
costs restricts a firm explanation, but perhaps the reason is that the size of building was much smaller in the provinces than in Rome and Italy.

Adding to this wide-scale construction of public buildings and structures was
the development of roads and bridges leading to and from Rome. In the whole em-

pire, approximately fifty-six thousand miles of paved highways were constructed by the second century A.D., and this network of highways was supplemented by more than two hundred thousand miles of secondary roads (Hopkins 1988). Nothing was spared, especially the landscape. The total depth of the roads, from the stone slabs at its top to its base, was about three to four and a half feet thick. Rivers were crossed with bridges made from wood or stone, mountains were pierced by tunnels, and embankments were made secure by retaining walls. For example, to the northeast of Rome, a road cut through the crater of an extinct volcano measured 1,625 yards long, 22 yards deep, and 20 feet wide (Hopkins 1988).

The administrative side of Rome that accounted for Rome's urban growth was spurred on by commercial and manufacturing activities (Laurence 1994; Finley 1977). These economic activities were tackled by labor housed in multistory blocks within the city or at the port of Ostia, which was close to the city. Textile production was the main economic focus, along with the production of pottery, glass, metals, and items of art and consumption.

Manufacturing and commerce resulted in a hierarchical stratification. Various occupational groups resided in the city and required housing; they thus contributed to the consumption of resources for the reproduction of luxurious lifestyles for those at the top of the social pyramid. According to Rome's tombstone records, there were at least two hundred named trades. A handbook indicates about 264 occupations that are comparable to the 350 occupational categories for London in the mid-eighteenth century A.D. (Hopkins 1978). Nearby Ostia also had numerous trade categories. Metalworkers accounted for 20 percent of its grave monuments; transport workers made up 17 percent (Randsborg 1991). There were also carpenters, cabinetmakers, leatherworkers, and shipwrights, which represented 10 to 15 percent of its grave monuments and markers. The remainder were mostly classified as food workers.

Besides the tradesmen, the stratification pyramid also included advocates, lawyers, doctors, architects, poets, playwrights, bankers, and schoolteachers. Clearly there was surplus generated that could support such diverse occupational categories. In fact, exuberant living was the order of the day; the rich elites ate from silver, and their wives bought jewelry and silk. Such luxurious spending even led to the passing of sumptuary laws to hold down expenditures.

Urbanization was not restricted to only the inner ring of the Roman world. The Roman lifestyles were reproduced throughout the republic and later the empire. Close to Rome lay Pompeii, with its population of about twenty thousand people. There, extensive textile manufacturing was carried out by laborers living in cramped two-story buildings. Moeller's study (1976) has shown that at least forty textile establishments operated in Pompeii, with some employing as many as twenty workers. Arezzo specialized in tableware, and by the first century A.D. it was a large industrial center. Its decline was taken over by La Graufesenque, located on the tributary of the Garonne. Diversity of production is reflected in lists indicating the production of seventy thousand different types of pottery items and the names of several thousand potters (Hopkins 1978).

Further from Rome but still within the confines of Italy, urbanization also flourished. Cities such as Bologna, Genoa, Modena, Parma, and Turin grew and were enhanced further by Roman colonists exported from Rome itself who were survivors of foreign wars or urban proletarians. The establishment of scattered Roman colonies was encouraged throughout the Mediterranean basin, and especially in the Roman provinces. The scale of such migrations was quite extensive; for example, Julius Caesar and Augustus arranged for the foundation of over one hundred Roman colonies. These colonies and garrison towns accelerated the urbanization trends. For instance, Cologne, Strasbourg, Vienna, Belgrade, and Mainz became major urban-growth centers fueled by the legions and colonists. Consumption thus went up and this further facilitated the urbanization process with the growth of workers and artisans involved in producing commodities for local consumption. These towns had quite substantial populations; some, such as Apamea, Ephesus, and Pergamum in the eastern Mediterranean, exceeded one hundred thousand residents. There were nine hundred towns in the eastern provinces, more than three hundred in North Africa, and also around three hundred in the Iberian peninsula and Italy. According to archaeological excavation, the built-up areas covered areas ranging from 200 hectares to more than 400 hectares. Rome itself is estimated to have been about 1,400 hectares (Randsborg 1991). Ephesus in Greece covered an area about 300 hectares. In Gaul, Nimes and Vienne were about 200 hectares each, while Alexandria, in Egypt, was more than 400 hectares. There were also towns of smaller sizes. Leptis Magna covered an area of 120 hectares, while Thugga was about 20 hectares, and Ostia and Pompeii each were in the 65-hectares range (Hopkins 1978).

The towns in the provinces and the frontier functioned in various ways. They were centers for economic exchange and manufacturing, and regional trade passed through them. They also served as places to supply the Roman legions garrisoned nearby and as administrative points of the empire. Such multiple functions facilitated the urbanizing trend.

As manufacturing concerns, these towns listed state factories established to supply the military. In addition, there was a high specialization of trade skills. Even small towns such as Cilicia in Asia Minor supported more than 110 trades, ranging from textile- and metalworkers to goldsmiths and foodworkers, by the third century A.D. (Hopkins 1988). Organization of these towns was based on the development of an urban community with lifestyles similar to that of Rome. The streets were laid in a rectilinear grid, and buildings at first were constructed in timber and toward the later period of the empire in stone, marble, and granite (Hanson 1988). Reflecting Roman tastes, mosaics were employed for the buildings; in the eastern provinces, such as in Anemurium, the palaestra's floor was covered by a mosaic to an area of larger than one thousand square meters (Randsborg 1991; Russell 1980). Statues adorned the streets in the eastern provinces, and large theaters and public baths were built in response to the Romanization process. The public baths had central heating, and the furnace, fueled by wood, pumped heat through the space between the floors. Amphitheaters were built; one with a seating capacity of between two and three thousand has been unearthed in Aquincum. Residential housing had underground drainage systems (Poczy 1980). In Roman North Africa, extensive waterworks were constructed, comprising aqueducts and reservoirs to meet the needs of the urbanized communities (Shaw 1995).

The garrisoning of Roman legions, especially in the outlying provinces, and the frontier of the empire exacerbated the transformation of the landscape. The building of a string of fortified encampments and fortresses to house the legions and as a line of defense further encouraged the development of nearby towns and cities, which supplied the forts with goods and services (Cornell 1995). In the east, the forts were built to accommodate about one thousand to fifteen hundred men (Parker 1984). They were rectangular in layout and about eleven acres in area. Craftsmen, traders, and the like lived in nearby urban enclaves to meet the needs of the military. Military needs were usually a priority and required the transportation of a large amount of commodities over distances with supply contracts amounting to 4.5 million denarii per year for the legions that have been recorded. The Roman legions also fostered the production of items in nearby towns to meet their needs (Randsborg 1991). For example, the manufacture of tiles and leather products was fostered and exchanged.

A fixed frontier with a string of forts was enacted by A.D. 1 (Randsborg 1991; Parker 1984). The building of forts became more frequent by the second half of the third century A.D., when Roman rule was increasingly being challenged at the frontier areas both in the west and east. Such was the case in the western part of the empire along the Rhine and in its hinterland as far as northernmost Gaul and along the upper Danube. Also in the east, where the Persians were making incursions, by the mid-third century A.D., the number of forts had more than doubled. Furthermore, to stave off these attacks, the linear line of defense was changed to one of "defense in depth" (Parker 1984). In the easternmost parts of the empire, this meant a deep frontier comprising forts, fortified towns, and a system of watchtowers along the rural areas, all linked by a network of roads. The width of this defense area was twenty to thirty kilometers, with the forts as anchors (Parker 1984). These forts had extensive associated civilian settlements nearby that either had been established earlier, before fort construction, or arose as a consequence of the garrisoning of troops. Ensuring imperium and Roman peace in this way consumed a tremendous amount of resources, from timber for building and as a source of fuel, to limestone, chert, and basalt for the walls. These were about 2.4 meters thick and were supplemented by projecting towers.

POPULATION

Clearly, the urbanization process that occurred was resource-intensive in nature. *Pari passu* to this process, the population levels in the urban areas were quite dense. Overall, the Roman empire had a population of between 50 and 60 million persons by A.D. 1 (Hopkins 1988; Russell 1958). In terms of population distribution, the European portion of the empire had close to 23 million persons, the Asian part had close to 20 million, and the African section had about 11.5 million persons (Russell 1958, 7). Average density was about 16,000 persons per square kilometer. The overall level is misleading as there were Roman provinces of extremely high density, such as in Asia, Syria, Cyprus, Egypt, and Cyrenaica. The average density in these areas was about 72,600 persons per square kilometer. For Roman Italy, however, the range covered from 4,000 persons per square kilometer to 38,000 persons per square kilometer during the period around A.D. 1. The latter figure is equivalent to modern-day Calcutta.

Rome, as indicated above, had a population of about 1,000,000, though some sources have estimated it at between 750,000 and 1,000,000 by this time period (Hopkins 1978; Storey 1997). Other cities also had quite large populations. Pergamum in the second century A.D. had a population of close to 200,000, and Antioch, Alexandria, and Carthage each had populations of closer to 300,000. In terms of density, these urban environments were similar to European cities such as London and Berlin in the nineteenth century (Duncan-Jones 1974). For the empire as a whole, the population level was equivalent to that of China for the same time period, each with about one-fifth of the world's population.

Such a high population level entailed a high measure of resource consumption. Goods and food items had to be transported to the cities. As centers of manufacturing, the urban environments were also the main sources of pollution of the air and the river systems. It is to these ecologically degradative trends that we now shift our focus.

ECOLOGICAL DEGRADATION

Urbanization, population increases, manufacturing, and trade during Roman times had severe consequences for the ecological landscape. The drastic degradation of the environment and the pollution that followed were not only restricted to areas within the Roman empire but extended to external areas. Thus, an ecologically degradative shadow was cast over a broad domain that covered Europe, the Mediterranean, and Asia Minor, including distant lands where Roman trade flowed. In other words, the Roman degradative footprint on the ecological landscape was expansive and intensive.

Air Pollution and Lead Poisoning

Population increases and manufacturing during the time of the Roman empire required large quantities of heavy metals to sustain the high standard of exuberant living (Nriagu 1983; Hong et al. 1994, 1996). About 80,000 to 100,000 metric tons of lead and 15,000 tons of copper per year were mined to sustain the Roman economy (Nriagu 1983). Besides these metals, others including zinc (10,000 tons per year) and mercury (2 tons per year) were required for manufacturing operations. Consumption of lead in Roman times was high: about 4,000 grams per person per year (Nriagu 1983). In comparison, present-day consumption is only about 1,500 grams on average per person per year.

The scale of such usage, especially for lead and copper, was prompted in part by the increasing use of coinage, and also of plumbing, glass manufacturing, and architectural construction including the building of ships. Such high levels of utilization rested on the intrinsic properties of the metals, and in the case of lead, on its corrosive resistance and formability, which made it ideal for civilian and military uses. Furthermore, its atomic configuration made it ideal for functioning as a opacifier or colorant in the manufacture of glassware and mirrors during Roman times. The metal imparted a toughness and brilliance that facilitated the manufacturing process and added quality to the finished product. Lead was also used in bronzes for statuary and coinage. For the latter, though forbidden by law, most Roman bronze coins contained between 1 and 30 percent lead. In construction and other architectural and engineering applications, lead was used to make anchors in which iron clams were utilized to secure stone blocks of buildings. It was also applied to line baths and cisterns. A covering of 1.5 centimeters thick was utilized for these purposes. Consumption levels of lead must have been enormous considering the 11 public baths, 856 smaller private baths, and 1,352 cisterns and fountains in the city of Rome in the fourth century A.D. (Forbes 1964). Lead was also used in piping and aqueducts for water distribution. It was thus utilized not only within Rome but in parts of the empire such as Gaul, Germany, Asia Minor, and England. Of the ninety-five large aqueducts in service then, nine were found in the city of Rome, and the rest in Asia Minor, Spain, Gaul, and Britain. These aqueducts reached heights of more than 180 feet, as the Pont du Gard in Provence has shown (Hughes 1994; Shaw 1995). Lead pipelines were evident everywhere. For example, a lead pipeline from Mount Pila used more than two thousand tons of lead. In Gaul around Beaumont, seven lead pipes of up to twenty-seven centimeters in diameter and requiring ten thousand tons of lead were constructed to carry water from the reservoir (Nriagu 1983). Cities in Asia Minor such as Sardis, Ephesos, Smyrna, Miletus, and Nysa also built plumbing aqueducts made of lead during the period of the Roman empire.

No doubt such intensive use of lead and copper generated substantial emissions from open-air furnaces; uncontrolled smelting of ores led to the pollution of the atmosphere at the local and regional levels, notwithstanding the poisoning of the human population (Nriagu 1983; Hong et al. 1994; Kempter et al. 1997). In fact, the pollution was not only restricted to the regional level; it has been discovered to have reached into the middle troposphere of the remote Arctic region, which is about 3,238 meters above sea level (Hong et al. 1994). According to ice-core and peat-core measurements, air-pollution levels composed of lead and copper particulates from the Roman era attained large-scale hemispheric levels (Hong et al. 1994, 1996; Kempter et al. 1997). Figures 5.5 and 5.6 exhibit the time series of lead and copper production levels from early times to the period of the Industrial Revolution in Europe. These trends suggest that the human contamination of the environment has been an ongoing exercise, at least with available data, as far back as Roman and Sung Dynasty times. The Roman empire generated severe pollution of the environment. Figures 5.5 and 5.6 show that the Roman era markedly increased lead and copper concentrations in the atmosphere, which had negligible concentrations prior to then (Hong et al. 1994). This polluting source was eliminated with the decline and fall of the Roman empire; subsequently, lead and copper pollution levels fell significantly and did not increase again at the hemispheric level until the Medieval period.

Besides air pollution, Roman degradative practices caused poisoning and exposure among its peoples and on the landscape, such as in pollution of the watersheds, where the mining and manufacturing of the metallic ores took place. Locations where these metals were mined, such as in Central Europe, Cyprus, and the Iberian peninsula, had landscapes littered with slag. Slag levels reached about 16 to 20 million tons in the Rio Tinto region, 6 to 7 million tons in the Tharsis mine region, and 3 to 4 million tons in smaller sites in Spain. Mine shafts of ten feet in diameter covered the landscape of Spain, such as at Orihuela and Mazarron. These shafts traveled from open-cast pits to the level of the deposits, with huge galleries connected to the ore vein. The mines at Laurium had more than two thousand vertical shafts and more than eighty-seven miles of tunnels (Hughes 1994, 118). Some of the mines had shafts dug to a depth of about eight hundred feet.

Lead poisoning of humans occurred extensively because of the widespread usage of lead in Roman life. Nriagu (1983) has even suggested that the level of lead poisoning suffered by Roman elites might be one of the causes for the fall of the Roman empire. Besides the exposure to high levels of lead by workers engaged in the mining and smelting of the metal, a number that reached about 140,000 persons per year, urban dwellers were also exposed to high levels from the various uses of lead in daily life (Emsley 1994). As indicated previously, lead was used in high amounts for construction purposes and in the manufacture of items for household or decorative uses. The most prevalent exposure was from drinking water that came from lead pipes, lead-based solders, lead vats, lead-lined cisterns, and lead-bronzed kettles. Lead concentration in drinking water in Asia Minor, Germany, and England was exacerbated further by the prevalence of soft water, which was more corrosive of lead plumbing. Extremely high levels of lead contamination could also be found in food and drink, coming from food-storage containers as well as food colorants and preservatives. For example, both the flavoring sauces *garum* and *liquamen* contained significant levels of lead, as they were produced in lead containers or lead-glazed vessels. Contamination was also high in wines and other beverages. The use of lead extended to food preservation and medicinal purposes. Lead sulfide was prescribed for the treatment of scars and diseases of the skin and as a component for an eyewash and hair wash (Nriagu 1983).

Figure 5.5 Lead Production

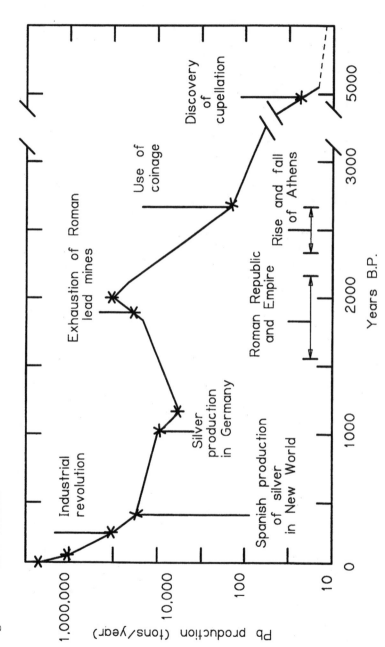

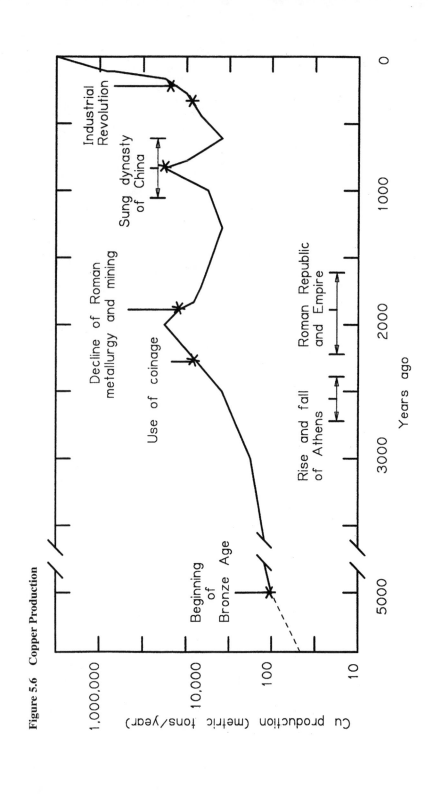

Figure 5.6 Copper Production

In view of the above, lead poisoning was prevalent all over the Roman empire. Table 5.1 shows the various sources and levels of lead poisoning across the hierarchical strata of Roman social structure. Daily exposure was quite high for the aristocrat, with a mean of 250 μg per day, in comparison with current recommended exposure levels of no more than 43 μg per day by the United Nations World Health Organization (Nriagu 1983). Persons at the lower levels of the social hierarchy suffered less exposure, primarily because of their more restrictive lifestyles and diet

Table 5.1   Exposure of the Subpopulations of the Roman Empire to Lead

Exposure Route	Pb Concentration in the Source*	Daily Intake Rate	Absorption Factor	Amount of Pb Absorbed (μg/day)*
Case I. The Aristocrat				
Air	0.05 μg/m^3	20 m^3	0.4	0.4
Water	50 (50–200) μg/liter	1.0 liter	0.1	5 (5–20)
Wines	300 (200–1,500) μg/liter	2.0 liters	0.3	180 (120–900)
Foods	0.2 (0.1–2.0) μg/g	3,000 g	0.1	60 (30–600)
Miscellaneous				5.0
Total				250 (160–1,520)
Case II. The Plebian				
Air	0.05 μg/m^3	20 m^3	0.4	0.4
Water	0.5 (0.5–5) μg/liter	2.0 liters	0.1	0.1 (0.1–1.0)
Wines	50 (50–400) μg/liter	1.0 liter	0.3	15 (15–120)
Foods	0.1 (0.1–1.0) μg/g	2,000 g	0.1	20 (20–200)
Total				35 (35–320)
Case III. The Slave (Labor Hand)				
Air	0.05 μg/m^3	20 m^3	0.4	0.4
Water	50 (50–200) μg/liter	2.0 liters	0.1	10 (10–40)
Wines	5 (1–10) μg/liter	0.75 liter	0.3	1.1 (0.2–2.0)
Foods	0.05 (0.05–0.5) μg/g	1,000g	0.1	5 (5.0–50)
Miscellaneous				5.0
Total				15 (15–77)

* Parentheses indicate range of intake.
(Source: *Lead and Lead Poisoning in Antiquity* by Jerome O. Nriagu, Table 6.3, p. 400. Copyright © 1983 by John Wiley & Sons, Inc. Reprinted by permission.)

in comparison with the exuberant living of the elites. Exposure to lead in the Roman empire was significantly higher than in other civilizations and empires prior to the eighteenth century. The calculated levels were an average of 2.4 μg per day for such other communities (National Academy of Sciences 1980).

Deforestation

Urbanization, high population levels, and accumulation brought about the increased extraction of heavy metals to reproduce exuberant lifestyles during the period of the Roman empire. As we have shown above, the effects were severe air pollution and lead poisoning. The Roman impact on the ecology was not restricted only to the atmosphere, for to reproduce the standard of living of the empire's urbanized communities and to meet the dynamics of capital accumulation, other aspects of the natural landscape were affected severely, such as the forests, the animal populations, and the soil.

Even as early as the second century B.C., forests were removed to give way to urbanized needs. As the city of Rome grew, houses began to replace the hillsides of Rome where forests once stood. To meet its timber needs at this stage, Rome subjugated Liguria and Umbria and conquered Etruria to provide timber to meet its building and ship-construction needs (Perlin 1989). Furthermore, the forests of the Po Valley also came under assault. As the power of Rome grew, expansion followed into Europe and North Africa. The forests initially suffered degradation primarily as a consequence of war campaigns. The size of the standing army was about three hundred thousand and increased to six hundred thousand toward the late empire period. Roman legions deforested areas where they camped or marched in order to reduce the cover where their adversaries could hide and/or mount a sneak attack. In his war campaigns in northern Gaul, Caesar ordered his legions to cut down the forests to prevent such sneak attacks (Perlin 1989). Large-scale wooden fortifications were built by the Roman legions, and wood was also used for the making of siege engines and for the transportation of supplies. Shipbuilding for the Roman navy also required a lot of wood. Large Roman navies were built during the war against Carthage.

Following Roman occupation, forested areas were reduced primarily for agriculture. Vast tracts of forests were removed in regions such as North Africa and other Roman provinces, in order to meet the grain needs of the urbanized communities of the Roman empire, and especially the cities in Roman Italy. For example, by the first century A.D., the Roman province of North Africa each year exported enough grain to feed 1,000,000 people for two-thirds of a year, or about 350,000 persons for a year (Meiggs 1982, 374; Murphey 1951, 123). Agricultural production and, in some provinces, the manufacturing of leather goods, textiles, and pots were further spurred on by the need to pay the allocated taxes to Rome (Hopkins 1978, 1980). Yields in some of the provinces were much higher than in Italy. For example, in Egypt the grain yields were ten times higher than in Italy (Hopkins 1978). Taxation forced the increase in productivity and also helped the growth of towns. Coupled with the imposition of money rent, the expansion of agricultural production was exacerbated further primarily through the demands of the landowners. All these tendencies spurred on deforestation.

Thus, forest loss was predominant all over the Roman empire by the end of the second century A.D.; this trend continued to the fourth century A.D., especially in the outlying provinces, where trees were cut down for mining, manufacturing, and agriculture (Van Vliet-Lanoe et al. 1992; Perlin 1989; Hughes 1975, 1982, 1983; Rosch 1990; Attenborough 1987; Randsborg 1991). For example, during the four hundred years of silver smelting in Iberia to meet the growth of Rome, five hundred million trees were cut (Perlin 1989, 125). Morocco lost 12.5 million acres of forests over the Roman period. Roman occupation of southern Germany caused a tremendous amount of deforestation from 50 B.C. to A.D. 150. Pollen analysis of this period for an area around Konstanz indicates the high degree of deforestation (Rosch 1990). Figure 5.7 provides an overview of a time series of deforestation that transcends the Roman period. It seems that there existed a cycle of deforestation and regeneration of the forests from at least 3000 B.C. for this region of southern Germany. This further suggests that the cycle of deforestation is not a particular feature of the Roman relationship with Nature. It is, in fact, a typical feature of human societal/civilizational relations with Nature.

The accumulation of wealth and urbanization coupled with population growth fostered exuberant lifestyles and a high standard of living for the urbanized communities of the Roman world. Temples, villas, and public buildings such as arenas, theaters, public baths, etc., were constructed to meet the varied needs of the elites all over the empire of Rome. The accumulation of wealth required expansion of trade, which was conducted via shipping. Urbanization required wood and lime-based concrete for the building of residential housing, furniture, household items, and aqueducts for the distribution of water supplies. Limestone kilning of plaster and mortar required a lot of wood. Roman kilns have been excavated at Tretau, Gaul, and at Iversheim, Germany (Hughes 1994). Additionally, extravagant architectural designs required intensive utilization of wood for balconies, ceilings, galleries, and roofs. At its zenith, Roman confidence led to exuberant lifestyles and rapid urbanization in the first century A.D., which generated the need for further utilization of wood to heat baths and villas, and to manufacture glass, pottery, bronze, and iron. To maintain temperatures of between 130 and 160 degrees Fahrenheit for a single public bath, 114 tons of wood were required per year, and for central heating of a Roman villa, more than two cords of wood per day were needed (Perlin 1989, 112). The mass appeal of glass for the manufacture of household items fostered an increase in glass manufacturing, which meant a heavy demand for wood as fuel production. When supplies of wood diminished due to continuous extensive deforestation, industries had to be relocated from Italy to other areas in Europe so that the fuel sources were closer to production. By the second half of the first century A.D., ceramic factories were established in southern France, and, in turn, these products were exported back to Rome (Chew

Figure 5.7   Pollen Count

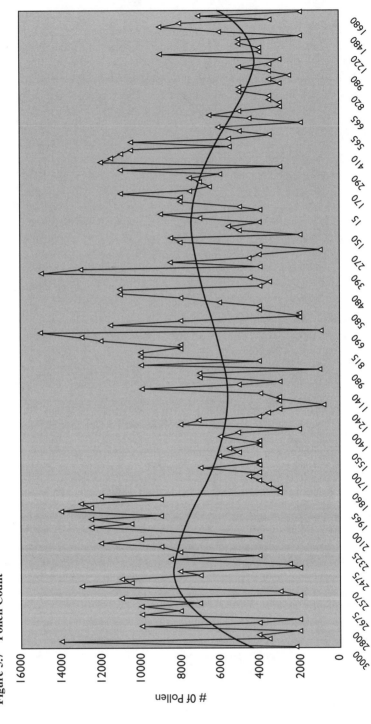

1997a). Glass and iron manufacturing were also shifted to regions such as Gaul and southern France. Later on, with increasing deforestation, the various bronze, glass, and pottery manufacturing processes established in southern France had to curtail their production levels. The iron mines in Britain and the copper mines in Cyprus also suffered due to the lack of adequate supplies of wood for operations.

Soil Erosion and Endangerment of Species

Deforestation of the landscape caused a tremendous amount of soil erosion. The Mediterranean ecosystem is quite mountainous and has a rainfall pattern that occurs mostly in winter. These features make the landscape vulnerable to soil erosion should heavy deforestation occur. The end results are flooding, disruption of water supply, and the siltation of coastal areas. As a consequence of the rapid and extensive deforestation during the Roman era, soil erosion was the order of the day. Erosion rates increased more than twentyfold from the second century B.C. onwards (Hughes 1982). Agricultural productivity suffered because of siltation and salinization. Siltation clogged canals and ditches, thereby increasing the amount of dissolved salts in irrigation waters. Besides this, the loss of topsoil made agricultural lands less productive. According to Hughes (1982, 69), "soil erosion has contributed to agricultural decline throughout much of the Mediterranean basin and seems to have reached a particularly critical stage during and after the third century A.D., when the classical cultures were in decline." Siltation also caused problems in terms of port closure. Clogged harbors at river mouths were a constant concern for the Romans. For example, the port of Ostia, which was Rome's coastal gateway in terms of trade, often experienced heavy siltation.

Soil erosion also caused sediments to collect in the low-lying areas, where marshlands formed as a result. The spread of malaria and other swamp-bred illnesses were predominant during the Roman period (Hughes 1994).

Besides the spread of diseases facilitated by the degradation of the environment, exuberant living also led to extinction of certain species. In this case, the Roman ecological degradative shadow extended to a wide area beyond Italy to Africa, Asia Minor, and beyond. Animals such as elephants, bulls, lions, and leopards were imported for entertainment in the arenas. Mock hunts, or *venationes*, were staged in the arenas for public entertainment. Emperor Augustus held twenty-six *venationes* in which thirty-five hundred animals were killed, including tigers from India (Hughes 1994). Emperor Titus at the dedication of the Roman Colosseum had nine thousand animals killed, while Emperor Trajan, to celebrate his conquest of Dacia, sacrificed eleven thousand wild animals. The epic level of gore left the sand of the arena soaked with blood following a *venatione*. *Venationes* were held not only in Rome but also in most towns throughout the empire. Such spectacles generated a high demand for wild animals, and commercial trading of them was an organized business.

The demand for these animals, which extended to birds as household pets or for their exotic colorful feathers, reached proportions that caused the elimination of some species from their habitats in Africa and Asia Minor. Elephants, rhinoceri, and zebras became extinct in North Africa. Lions were made extinct in western Asia Minor and nearly so in the Atlas mountains. Tigers were made extinct in Armenia and in northern Iran, which two areas were the closest source for Rome. Bird populations were also diminished, such as flamingos, which used to flock the Rhone River. In addition to the import of these wild animals for the arenas, hunting also extirpated a number of species, as did the systematic destruction of their habitats through agricultural expansion.

Ecological Conservation

Despite the fact that the Romans' relationship with Nature could be characterized as exploitative (as Cicero so vividly puts it, "We are the absolute masters of what the earth produces. We enjoy the mountains and the plains, the rivers are ours. We sow the seed and plant the trees. We fertilize the earth. . . . We stop, direct, and turn the rivers, in short by our hands we endeavor, by our various operations in this world, to make, as it were, another nature" [quoted in Hughes and Thirgood 1982, 206]), there were also occasional attempts to conserve and recycle natural resources, especially when these resources were dwindling.

Wide-scale deforestation was coupled in places with afforestation, wherein landowners transplanted cuttings and young trees. Some farms did not clear all the trees on the farm but kept a woodlot, and on some large farms forests were retained to meet the wood needs of the farm. For the elites, there were also attempts to maintain artificial parks of exotic trees or trees of many species planted in a disorderly form to create a human-generated wilderness (Hughes 1982, 1994). In places where trees were in high demand and where supplies were dwindling, such as in Egypt, tree plantations were established on private and royal lands. These plantations had forest regulations that prohibited entry to animals such as sheep and goats, and laws stipulating the age and size of trees that were available for logging or thinning.

Conservation was also enhanced in places where groves of trees and forests were protected through religious identification as sacred areas. Sacred groves had rules that prohibited logging and the entry of animals other than for ritual sacrifices. Penalties were imposed by local magistrates for desecrating scared groves, ranging from whipping for slaves or aliens to severe fines for citizens.

Beyond this, there were also government regulations that supervised the use of the forests and watersheds. Regulations governing the construction of buildings and other works that were to be built in forested areas were introduced. A forest guard service was established to ensure compliance so that forest resources could be protected for use in production processes (Perlin 1989).

Dwindling wood supply also forced recycling in areas where wood as fuel was in high demand. Glasswares were recycled to minimize wood consumption. Culinary experts suggested various additions to cooking that would reduce cooking times. For example, adding stalks of wild figs to beef was thought to reduce cooking time and maintain tenderness of the meat. As with the Greeks, Roman architectural design utilized the concept of solar heating whereby buildings were aligned with southern exposures. Another architectural feature was the extensive use of glass as solar-heat traps in the windows of villas (Butti and Perlin 1980, 13).

Such was the state in which we find the ecological conditions and relations in the Roman empire. As we continue our journey through world history, do these trends and tendencies underlying human relations with Nature continue to be repeated over time and space? It is to more recent world history that we turn in our next chapter.

Chapter Six

The Sun Rises and Sets in the East: Trade, Hegemony, and Deforestation A.D. 500–A.D. 1800

The trading world of Asia in A.D. 500 spanned across the Indian Ocean and the Bay of Bengal to the South China Sea. It was a world that was circumscribed by developed communities, with India on the western borders and China anchoring the east. In between India and China were communities and kingdoms located on the Kra peninsula, Indochina, the Malayan archipelago, and the Indonesian islands. This trading world extended itself via land and sea with connections to the East African coast, the Mediterranean, and beyond. It was a world where manufactures, preciosities, natural resources, grains, animals, and even human slaves were exchanged. Surpluses in the form of gold and silver were amassed, for these developed communities produced items that were sought in the Mediterranean and beyond. Accumulation and exuberant consumption were the order of the day.

With these economic and materialistic tendencies, extensive utilization of natural resources, land clearing for agriculture (wet rice, pepper, etc.) and animal husbandry, deforestation was the normal outcome. As with the other civilizations and kingdoms that we have examined, the repercussions of these human activities extended beyond the realms of the civilizations and kingdoms of the trading world of Asia. In addition to this materialist reproduction of sociocultural life, population increases and wars were other dimensions that impacted and transformed Nature–Culture relations.

One of the most visible outcomes of the excessiveness of cultural domination of Nature is the process of deforestation, which can serve for us as a proxy for revealing the level of ecological degradation of the environment, and the level of consumption and accumulation for this period of world history.

THE EAST AS CORE

The Trading World of Asia

For Asia, the core centers of accumulation were located in China, South Asia, and the city-states of Southeast Asia. With the sea route paralleling the Silk route of Central Asia to the Mediterranean and the West, the zone circumscribing the Red/Arabian Seas, the Bay of Bengal, the Straits of Malacca, and the South China Sea was one of trade exchanges linking cities and kingdoms located in southern Arabia, southern India, the Malayan archipelago, Sumatra, Java, Indo China, and southern China with the Mediterranean (Wang 1958; Wheatley 1961; Hall 1985).[1] These trading routes were distinguished by the different merchants and traders that plied the waters. For the Red and Arabian Seas and the Mediterranean, the trading relationship was dominated by Arab and Greco-Egyptian merchants. The Bay of Bengal region was controlled by Indian merchants, and further east around the Malay peninsula to the shores of Indochina the exchange of merchandise was undertaken mainly by Malay merchants, with Yueh traders and sailors dominating the Gulf of Tonkin and south China trade (Wheatley 1964b, 41; Hall 1985).

With the prevailing monsoon winds that blew in various directions over this zone during different times of the year, the merchant ships could sail to different ports depending on the season of the year (Chaudhuri 1985; Abu-Lughod 1989; Wheatley 1961; Wang 1958; Hall 1985). The northeast and southwest monsoon winds would bring one set of merchants from one part of this zone to Southeast Asia, and at the same time take another set of merchants home. For example, when Indian and Arab ships arrived in Southeast Asia on the southwest monsoon, the Chinese junks would be sailing home.

Early Southeast Asia

The Nature–Culture relations in Southeast Asia started quite early in the course of world history. Wet rice cultivation occurred as early as the third millennium B.C., with evidence of burning in the lower Bang Pakong Valley in central Thailand as early as the fifth millennium B.C. (Higham 1996, 309). Neolithic communities have been unearthed at Khok Phanom Di, where a widespread exchange network existed as early as the Bronze Age with bronze being forged (Higham 1996; Bayard 1992; K. Taylor 1992; Hall 1992; Cresmachi and Piggott 1992). Soil erosion could be seen as early as 1000 B.C. in the Khao Wong Pra Chan valley in Thailand as a consequence of deforestation to meet fuel requirements that were an outcome of the smelting processes (Cresmachi and Piggott 1992, 196). Reforestation took some time as the soil was impregnated with high concentrations of copper.

The previous chapters have underlined the trade exchange between India, China, Mesopotamia, and the Mediterranean as far back as the third millennium

B.C. Timber and wood products were sought after, especially the scented woods, in the urbanized communities of southern Arabia, southern India, and southern China. The first assault on the forest cover of India started as early as the third millennium B.C. and in Southeast Asia (such as the Malayan peninsula) about 2500 B.C. (Erdosy 1998; Wheatley 1961; Tibbetts 1956). Tibbetts (1956, 183–84) has stated that there was evidence of Indian wood found in early Sumerian cities and on inscriptions, indicating the importation of wood from India as early as 2000 B.C. Northwestern India had forests but was degraded by the Harappan civilization as described in chapter 2, though according to Erdosy (1998, 60), whether extensive forests ever existed in northwestern India remains a mystery.

Bellwood (1985) has suggested that by the middle of the third millennium B.C., there was expansion of pottery use and agriculture in Southeast Asia.[2] There were also localized exchange networks in Indonesia and Malaya from the second millennium B.C. onwards (Glover 1979, 1996). Through a natural course of development, Southeast Asian merchants and trading communities were already participants in the trading world by 1000 B.C. and had commercial contacts with India by this time period (Leong 1990, 20–21; Christie 1990; Hall 1985). From 500 B.C. onwards, large moated settlements existed in central Thailand, where iron weapons and tools have been uncovered (Higham 1996; Watson 1992). Evidence of salt extraction has also been unearthed. All these materialistic activities underscore the fact that the forests of Southeast Asia were being removed to meet these economic tendencies. Evidence in terms of volume of wood extracted is lacking, other than the qualitative comments by Wheatley (1961), Dunn (1975), Leong (1990), and Gadgil and Guha (1992). We begin to obtain a better idea of this trade in wood products after the first century A.D., especially the trade between the kingdoms and city-states in Southeast Asia and China.

What is clear is that by the beginning of the first century A.D., trade flourished among India, the Far East, and the Mediterranean (Tibbetts 1956; Colless 1969; Abu-Lughod 1989; Christie 1990; Hall 1985). Besides luxuries and spices, other products traded were timber, brazilwood, cotton cloth, swords, sandalwood, camphor, rugs, metals, and even African slaves (Wheatley 1959; Lim 1992; Lian 1988; Leong 1990; Christie 1990; Hall 1985). By the first century A.D., Malay sailors were settling along the East African coast (K. Taylor 1992; Hall 1985). Glover (1996) and Hall (1985) have also asserted that by this time period, Southeast Asia was already part of a world trading system linking the civilizations of the Mediterranean basin and Han China.

Records reflect the trade in wood products between China and Southeast Asia (including Japan), with the former supplying manufactured products and the latter natural resources, of which wood was one of the primary commodities (Wheatley 1959, 1961; Wang 1958; Wolters 1967; Coedes 1983; Yamamoto 1981; Totman 1989; Leong 1990; Hall 1985). The lines of commerce were quite well established before the first century A.D., and timber and other wood products were part of this trade exchange (Solheim 1992; Hall 1985).

Wheatley (1964a, 41) has noted that by the third century A.D., peninsular Malaya was undergoing radical socioeconomic changes. These occurred primarily because Indian merchants and traders were exchanging their merchandise and wares along the coastal areas of Southeast Asia, seeking gold that in the past they had obtained from the Mediterranean. With the prohibition on the export of gold imposed by Roman Emperor Vespasian (A.D. 69–79) this spurred the Indian merchants to search for gold bullion in Southeast Asia (Hall 1985). Indian ships weighing about seventy-five tons and that could carry up to two hundred persons were sailing between South Asia/Ceylon and China by the beginning of the Christian era (Wheatley 1964b, 34). Such trading patterns continued into the Christian period, and by the end of the ninth century A.D., Southeast Asia was known to a large number of Arab, Indian, and Persian traders and the goods from this area were taken back to the Middle East (Tibbetts 1957; Colless 1969; Dunn 1975; Hall 1985). The height of economic expansion of this trading world occurred during the period of the Tang and Sung Dynasties, from the seventh century A.D. onwards, when city-states located in the Malay archipelago, Annam, Java, and Sumatra enjoyed trade exchanges (including wood products) with China. These trading relations continued into the fifteenth century, prior to the arrival of the Europeans.

From the first to the third century A.D., the trade in wood products grew in Southeast Asia. Gharuwood was imported to southern China, involving merchants from the Malay archipelago, Sumatra, and even as far as Ceylon (Hall 1985). City-states such as Lo-yueh (near Hanoi) were the collection centers for forest products, while P'eng-feng produced lakawood (Wheatley 1961, 60, 71; Hall 1985). Further south, Tun-sun, situated on the Malay peninsula, was a dependency of the state of Fu-nan in Indochina, which was the core kingdom in the region.[3] Tun-sun was the emporium that lay on the great trading route between the Mediterranean and eastern Asia (Wheatley 1964a, 46; Hall 1985). It had five princes who were vassals of Fu-nan. Besides Tun-Sun, there were other kingdoms in Southeast Asia: Tung Tien, Ch'u-tu-k'un, Chiu-Chih, T'ai-p'ing, Yu-lan, Sui-Shu, Tu-k'un, Pien/Pan-tou, Pi-Sung, Chin-lin, and Chu-li (Wheatley 1964a). Evidence indicates that large urban communities were the order of the day. For example, Tung Tien in the third century A.D. had over twenty thousand families, which would give it a population of eighty to a hundred thousand persons.

Judging from the amount of tribute provided to China, these states must have been prosperous and economically developed. Lo-yueh, for example, was said to have twenty thousand soldiers and numerous palaces. Fu-nan was a major core center of accumulation from the first century A.D. to the sixth century A.D. (Stark 1996; Hall 1985, 1992). It was ruled by the Kaundinya clan in A.D. 1, and this rule was transferred to the dynasty of Fan by A.D. 2. By the third century A.D., Fu-nan consolidated all of the trading marts in the Malayan archipelago; its control was extensive and ranged from the Indian Ocean to the South China Sea (Wheatley 1964b, 45; Hall 1985, 1992). It became the international transit center for trade from Arabia, India, and China. Frankincense, myrrh, camphor, spices, gharuwood, and sandalwood were transshipped from Southeast Asian sources for exchange in the ports of southern China for Chinese silks, which were then shipped westwards to India, Arabia, and the Mediterranean. Within this trading world, the economic and political status of Fu-nan is a core one if we examine its relations with its immediate neighbors—like Tun-sun, for example—but when placed within the context of its relations with China it had, perhaps, a semiperiphery status (Stark 1996). Roman coins and products from Iran have been discovered among the ruins of Fu-nan, indicating the extent of the trading exchange. Besides this, Wheatley (1964a) has also indicated strong Iranian influence in Fu-nan. Wood, tin, and aromatics were exported to India.

Moated settlements represented urbanization and were located between upland areas and the alluvial plains. According to Stark (1996, 11), the moated urbanized communities sought upland resources such as timber, salt, and laterite ores for smelting. The mode of utilization of ecological resources was similar to that of other communities examined in the previous chapters. There also occurred household production of commodities that sustained systems of community-based specialization. Such political hegemony lasted only about half a millennium, for by the late sixth century A.D., with the fall of Fu-nan, the kingdom of Ch'ih-t'u (situated on the Isthmus of Kra) became a substantial power in the region (Wheatley 1961; Wang 1958; Dunn 1975).

Srivijaya

With the fall of Fu-nan, by the seventh century A.D. we witness the rise of Srivijaya and the part it played in the Asian trade (Wolters 1967; K. Taylor 1992). From the seventh century A.D., the kingdom of Srivijaya, situated in southeastern Sumatra, developed into a regional power, and it played an important role in the region for the next two centuries (eighth and ninth) (Rahman 1990; Wolters 1967, 1970; Wang 1958; Wheatley 1961; Coedes 1983).[4] Indonesian trade with the rest of Asia by no means was initiated with the rise of the kingdom of Srivijaya. As early as the third century A.D., horses were imported from northwestern India, and gold may have been exported from the Indonesian islands to India when the usual source of gold for India was curtailed in the early part of the Christian period (Wolters 1967). In addition to the commercial exchanges between Indian and Indonesian merchants, early Indian texts also note the importation of gharuwood and sandalwood from Indonesia to India, along with spices such as cloves, peppers, etc.

As noted earlier, tree products were exported from Southeast Asian regions such as Fu-nan and Tun-sun in Indochina and the Malayan archipelago (Leong 1990). Besides being derived from the hinterlands of these places, these products were also derived from the Indonesian islands (Wolters 1967; Leong 1990). By the fifth century A.D., the growing trade connections via the sea between China and Sassanid Persia further increased the importance of the Straits of Malacca to this maritime commerce. This rise was a consequence of the growing size of the market and economy of southern China, which developed from the massive flight of the Chinese southwards when incursions were made in northern China. The fall of Lo-yang in the early fourth century signaled a major shift in the migration of the Chinese upper classes to the south of the Yangtze. Naturally, the Indonesian merchants benefited from this increased pace of maritime trade.

In addition to gharuwood and sandalwood, pine resin was a major item traded between the Indonesian islands and southern China. By A.D. 500, benzoin was also a major item of import into southern China from Indonesia. The product primarily served a fumigatory purpose. It became the substitute for myrrh from Arabia and Africa. According to Wolters (1967, 111), benzoin grew into a valuable trading commodity not only in China but in western Asia and Europe after the Srivijayan empire developed its trading relations, starting in the seventh century A.D. onwards.

The consolidation of power by the Srivijayan kingdom through naval power can be seen in its turning Kedah, based in northwestern Malaya, into a dependency (Wolters 1967; Rahman 1990). Other candidates for dependency included Kataha and Chaiya. To enforce this political hegemony, the kingdom had an armed force of close to twenty thousand soldiers. For trade and naval power, ships of 250 to 500 tons measuring fifty meters long were built. These ships had multiple masts and sails (Manguin 1985, 300). No doubt a significant number of trees was required to meet the shipbuilding process, as the ships were of the sewn-plank type. Such consolidation of control of the kingdoms and ports on the Sumatran and Malayan coasts led to the kingdom's becoming the main export and collection point for the products of Southeast Asia and an entrepôt center for international trade with Arabia, the Mediterranean, India, and China. Its power started to decline from the eleventh century A.D. onwards, when it was at war with the Cola empire of southern India and the Javanese. Kedah, once a dependency, by then had lost that status and began also to challenge the Srivijayan kingdom. The end of the eleventh century marked the end of Srivijaya's dominance. Its capital by then had shifted to Jambi; in the past it had been at Palembang.

The passing of two centuries, with the rise of the kingdom of Majapahit in the late thirteenth century A.D., saw the continuance of Southeast Asians in the Asian trading system that before had been dominated by the Srivijayans. Located in Java, the Majapahit kingdom flexed its political hegemony by sending its navy into the Java Sea, which was the main trading route for the spices and forest products coming from other Indonesian islands. This projection of power was also extended to other parts of Southeast Asia, though according to Keith Taylor (1992) it was more symbolic than actual. At most, the Majapahits could claim control over Java, Bali, Madura, Borneo, Celebes, and Sumbawa, and also for a time the Straits of Malacca. The latter locale was challenged by Malacca, which increasingly became prominent as a trading port by the end of the fourteenth century

(Hall 1985). Majapahit control was short lived, and by the end of the fifteenth century A.D. it had ended, leaving a power vacuum that the early Europeans exploited when the first Portuguese fleet sailed past the Cape of Good Hope into the region in the early parts of the sixteenth century.

Deforestation in Early Southeast Asia

It is clear from what has been discussed thus far that significant civilizations and kingdoms existed in Southeast Asia prior to the arrival of the Europeans. The undertaking of agriculture and the manufacturing of items such as beads, pottery, etc. indicates the utilization of natural resources to meet economic and social-reproductive activities. Especially in the area of agriculture, wet-rice cultivation, for example, meant the removal of forests. The earliest centers for this type of agriculture lay in the intramontane basins and sloping foothills of Burma, Siam, Laos, Vietnam, Java, and Bali. By eighth century A.D. intensive wet-rice cultivation had taken root in Burma. The existence of complex urban communities indicates surplus production of rice that must have been shipped to these urban locales. Increases in population naturally required further production surpluses that led to further forest removal. Though the population levels were not as dense as those of India and China during this period, Southeast Asia by 1600 A.D. had more than 23 million persons (Reid 1992). Vietnam, Burma, Java, Sumatra, and Siam were the major areas, with population levels ranging from 1.7 million to 4.7 million persons. Spurred by clearing for agriculture and by the utilization of wood for fuel in manufacturing, deforestation occurred.

With deforestation occurring as a consequence of the rising trends in accumulation, urbanization, and population levels, the harvesting of forest products for export further heightened the human relationship with Nature (Boomgaard 1998, 379). Reid (1998, 106) has argued, however, that the increase in land use did not arise until the fifteenth century, when the rapid growth of commercial agriculture—specifically, the growing of pepper, sugar, cloves, gambier, and coffee—had resulted in the deforestation of large areas in Sumatra, Java, Borneo, Vietnam, and the Malayan Peninsula. Without sufficient data, such as pollen analysis, it is difficult to examine critically the assertion that the Southeast Asians were living in an interdependent harmonious relationship with the land prior to the fifteenth century. From what little empirical information is available—such as that presented by Wang (1958), Dunn (1975), and Wheatley (1959) about the volume of forest products shipped to China and the gold and silver sent as tribute (indicated previously and later in this chapter)—it can be surmised that the relationship cannot be described as an interdependent one that did not exploit the environment. No doubt, this exploitation intensified after the fifteenth century A.D.; by no means, however, can we consider Southeast Asian civilizations and kingdoms the exception to the rule in terms of their relations with Nature in comparison to the other civilizations and kingdoms that we have surveyed so far.

After the fifteenth century, the increasing use of the land to meet an export market and to satisfy the demands of the growth in population caused further deforestation. Cloves and nutmeg plantations were cultivated in Southeast Asia (Pires 1944). Pepper and cotton, important exports of Southeast Asia, were also grown. The commercial scale of agriculture led to intensive deforestation near where these crops were cultivated. According to Reid (1998, 113), the exported volumes of cloves and nutmeg to the Mediterranean ports reached about thirty tons and ten tons per year respectively by A.D. 1400. By A.D. 1600, this rose to over two hundred tons and one hundred tons respectively. Pepper and sappan-wood, which in the past had been considered exotic luxuries, by the 1400s were considered consumer items in China and were imported regularly (Reid 1992; Finlay 1992). By A.D. 1600, the volume of pepper exported from the Indonesian islands rose to over five thousand tons a year. This meant that extensive defor-estation occurred in northern Maluku, Ambon, the Banda archipelago, Sumatra, west Java, the Malayan Peninsula, and Borneo. In the case of Sumatra, pepper-growing in the seventeenth century resulted in the felling of over 120,000 hectares of forest; this area never returned to forest but instead turned into grass-land (Reid 1998, 116). The deforested area increased in the following century to 165,000 hectares and climaxed at 475,000 hectares. Cotton-growing areas such as eastern Java, Bali, Limbic, Samba, and Baton were also deforested.

Cotton and sugarcane cultivation also generated deforestation. The forests on the islands of the Philippines, such as Lauzon, Cebu, and Panay, were removed to grow cotton. On the Southeast Asian mainland, parts of Burma along the Ir-rawaddy River were also deforested to grow cotton. When sugarcane was planted in central Vietnam, Siam, Cambodia, and Java, deforestation occurred there.

Included in the cultivation of cash crops was benzoin, which was grown in large plantations in Sumatra, Siam, Cambodia, and Laos. Yearly output was quite high; for example, the Mekong valley exported about 270 tons per year by 1630. Deforestation thus was the order of the day.

China

Kennedy (1898) has suggested that trade between Mesopotamia and China began as early as the seventh century B.C. (Tibbetts 1956). According to Wang (1958, 21), trade occurred between China and ports on the Indian Ocean by the second half of the first century B.C. Others, such as Wheatley (1959, 19), have reported that Chinese envoys were sent by the Han emperor Wu (141–87 B.C.) to explore the South Seas as far as the Bay of Bengal.

China's trading relations with Southeast Asia began following the unification of China in 221 B.C., when the Chinese pursued expansion to the south. By the second century A.D., the total number of people residing in the Tonkin delta came to about 1,372,290 (Wang 1958, 18). Tribute missions from the city-states and kingdoms of Southeast Asia, and also from south India, started arriving in China during the second century A.D. The purpose of such missions, according to Wang (1958, 119; 1989), was to pay tribute, including wood products, so that political and economic concessions could be obtained.[5] For example, the state of Lin-yi in A.D. 433 provided tribute to obtain territorial concessions in Chiao-chou, and the state of Funan in 484 A.D. demanded justice from the incursions of Lin-yi. Besides wood, gold, silver, and copper were provided to the Chinese court. A mission from Lin-yi brought tribute of ten thousand kati of gold, one hundred thousand kati of silver, and three hundred thousand kati of copper (Wang 1958, 52). Such tribute missions increased in volume as time progressed; by the era of the Tang Dynasty, a total of sixty-four missions was recorded (Wang 1958, 122–23).

These missions came as far as Sumatra, Java, and Pagan (Wang 1989, 1116; Hall 1985). China's relations with these areas changed over the rise and fall of the Chinese dynasties. The trading relationship between the kingdoms and city-states of Southeast Asia and China was buttressed further with the arrival of the Persian and Arab merchants in China by the seventh century A.D.; their activity occurred mainly in Canton, Hanoi, Yang-chou, and Ch'uan-chou.[6] Wheatley (1959, 28) has noted that these were mainly Middle Eastern Muslims who had their own mosques and suks established in these areas and who were governed by their own sheikhs. The power and number of these merchants grew; by the mid-eighth century A.D., they were of such substantial strength that they settled their disagreements with the Chinese by burning buildings and *godowns* (warehouses) in Canton in A.D. 758 Besides Southeast Asia, China also had trading relations with Korea and Japan; traders from these countries arrived in ports of the Hangchow region (Wang 1989, 1117).

The volume of trade continued and by the middle of the Tang Dynasty, the ships of Ceylon—over two hundred feet long and carrying six to seven hundred persons—were plying the waters of the South China Sea. These ships were probably built in China because of the abundant timber resources in the coastal areas of Chang-chou and in Ch'ao-chou, Hsun-chou, Lui-chou, and Chin-chou of the Kwangtung province (Wheatley 1958, 109). With the end of the Tang Dynasty, by A.D. ninth century, the Chinese could not command much tribute from the Southeast Asian kingdoms. The Vietnamese fought successfully for their independence, and it took the Northern Sung Dynasty, according to Wang (1989, 1118), "more than a century to accept [that] this change (i.e. Vietnam's independence) was not a temporary one." With this shift in relationship, Southeast Asia came to be viewed by the Chinese court as an area of only economic benefit and beyond the political hegemony of China, though there were efforts to regain authority following the installation of the Yuan Dynasty. By A.D. 987, during the Sung Dynasty, the southern maritime trade provided a fifth of the total cash revenue of the state; such a high volume led the state to support missions overseas to induce foreign traders to come to trade at Chinese ports (Wheatley 1959, 24; Hall 1985).[7] Various land wars were also fought in the areas surrounding Yunnan with the Tai peoples, who occupied Laos, Burma, and Thailand from the mid-eleventh century onwards. Expeditions were also sent to challenge Champa, Java, and Japan in the eleventh century A.D. In short, the effort was made to ensure that the maritime trade among China, Southeast Asia, and South Asia remained protected.

After A.D. 1127, China's overland Central Asian trade routes were cut off, and China proceeded to further exploit the sea route of the South China Sea (Wheatley 1961, 61). Ebony, gharuwood, lakawood, pandan matting, cardamom, ivory, rhino horns, and beeswax were imported from Asia and India (Dunn 1975; Wheatley 1959; Hall 1985). To illustrate the increase in trade, between 1049 and 1053, the annual import into China of tusks, rhinoceros horns, pearls, aromatics, and incense was about fifty-three thousand units; after 1175 it reached half a million units. The increase in activity naturally led to the emergence of a powerful merchant group, which gradually came to manage all the major governmental monopolies of this trade. With the Mongol control of south China in 1277, trade with the rest of Asia and the Middle East was further encouraged. Southern China by the end of the thirteenth century had about 85 to 90 percent of the country's population. The country experienced an expansionary phase between the ninth and fourteenth centuries whereby industry intensified and agricultural production went up. This must have led to deforestation to clear land for farming and to obtain wood to fuel the kilns for the manufacture of pottery and for metalworking. Metallic currency was used as payment for the products of Asia and the Middle East; additionally, pottery was exported. According to Wheatley (1959) and Yamamoto (1981), by the time of the Sung dynasty there was a deficit in the Nanhai trade, which was covered with payment by bullion and coins. During this period, China also experienced a metallic-coin shortage (Yamamoto 1981, 24).

After the fall of the Yuan Dynasty, the Ming Dynasty proceeded to build on the Nanhai trade. The Ming Dynasty under Emperor Hung-wu pursued the type of relationship that the Tang, Sung, and Yuan Dynasties had enjoyed with Southeast Asia. Tribute missions were encouraged from the kingdoms of Southeast Asia, helped along with the Ming emperor's dispatch of gifts. There was, by contrast, no demand for submission to China, as the Yuan Dynasty had pursued. Overall, however, the Ming Dynasty was not as supportive of the trade as the Yuan Dynasty had been (Wang 1989, 1126–27).

China had a sizable navy; by the end of the fourteenth century it had thirty-five hundred ships, of which more than seventeen hundred were warships and four hundred were armed transports (Lo 1958). From the type of wood imported, we can surmise that China utilized its own pine and cedar forests for shipbuilding; it could obtain these woods from the coastal areas of southern China. Between 1403 and 1433, several expeditions were sent, each comprising no less than three hundred vessels carrying twenty-eight thousand men in one fleet (Finlay 1992). The vessels were of considerable size, ranging from fifteen hundred to three thousand tons and measuring more than four hundred feet in length; each had about nine masts and carried over six hundred men. Under the command of Admiral Cheng

Ho, these expeditions sailed as far as Mecca, Aden, Mogadishu, and Juba (Su 1967). When these expeditions ended in 1435, the Chinese fleet was withdrawn from the Indian ocean, leaving a power vacuum that was exploited by the Portuguese later in the century.[8]

Deforestation in Early China

As a core center within the trading world of Asia, China's geographic size and political power meant that it could not only rely on its own ecological resources to maintain its accumulative practices, but could as well rely on other regions to meet its consumptive needs. This means that the trends and tendencies of deforestation occurring in China must also be coupled with what was occurring in Southeast Asia and beyond, regions with which China enjoyed trading relationships. Thus, what was discussed previously regarding deforestation in early Southeast Asia should partly be attributed to the socioeconomic demands of China at that point in time. Deforestation in China during its early history is a process similar to what we have witnessed previously in other civilizations and kingdoms. Though not occurring relentlessly throughout Chinese history, deforestation was an outcome of the needs and priorities of a Chinese society conditioned by accumulation, urbanization, and population increases. Like the ancient civilizations of Mesopotamia and the Indus Valley, rapid deforestation in early China for agricultural production, irrigation, and the production of metallic implements and commodities generated the conditions of ecological crisis (Teng 1927). With population increases requiring more arable land, and the increasing utilization of wood to provide palaces, temples, tombs, and other personal pleasures of the elites, more pressure was added to the already fragile ecological conditions (Teng 1927). Such conditions began to be felt from the fifth century B.C. onwards (Bilsky 1980). The river valleys of the Hwang Ho suffered severe deforestation during the Chin and Han dynasties (221 B.C.–A.D. 220). They were deforested to meet the urbanization needs of the capital cities of these dynasties.

The population increase over Chinese history does show an intensification of land use that resulted in rapid deforestation, which in the long run became permanent. Population increases in south China from the seventh to the eleventh century A.D.—associated with advances in agricultural technology such as the use of the plow, the techniques of transplanting and double cropping, hydrological advances, new seed varieties, etc.—all led to further deforestation (Reid 1998; Elvin 1993; McNeill 1998; Menzies 1992, 1996; Teng 1927).

From a population of more than 50 million before the end of the first millennium B.C., China grew to more than 100 million by 1000 A.D. The number jumped to 330 million by the eighteenth century A.D. and doubled again by A.D. 1850 (Elvin 1993; Durand 1960). Such increases, along with the urbanization process, led to intensive agricultural cultivation, and in the later period also led to land reclamation to increase food productivity (Osborne 1998; Marks 1996, 1998;

Pomeranz 1994; Teng 1927). The process of urbanization started as early as the Sung Dynasty (A.D. 960–1279) and continued until the eighteenth century A.D., though it emerged much earlier on a national scale—albeit with regional differences after the Tang Dynasty—and continued until the Sung Dynasty and beyond (Hartwell 1982).

The economic expansion during the Ming Dynasty also involved land reclamation to meet the agricultural demands of a growing population in the Yangtze delta. The filled landscape in the lowland areas prompted a shift to economic transformation of the highland areas. The transformation had a cost: the extraction of natural resources and the demands of rising prices led to degradation of the landscape without regeneration of the forests. The overcutting also led to a shortage of fuel, so that firewood sold at prices like those of precious cassia wood (Osborne 1998). The establishment of tea plantations and of other cash crops such as tobacco, indigo, hemp, etc. also led to deforestation and soil erosion—issues that have occurred transhistorically and transculturally.

According to Elvin (1993, 29), by the end of the imperial period (A.D.1900), most of China proper had been stripped of the forest cover that three millennia ago had produced an almost complete succession of canopies stretching from the tropical rainforests in south China to the conifers of the north. The most affected areas were the provinces of Fujian, Hunan, and Sichuan, which were the main forest-resource areas for China, especially during the Ming and Ching Dynasties (Adshead 1974; Marks 1998). According to Teng (1927), deforestation occurred in several phases. The first occurred from 2700 B.C. to 1100 B.C., followed by a period of active conservation from 1100 B.C. to 250 B.C. After this, from 250 B.C. until the twentieth century A.D., occurred a phase of severe deforestation caused by agricultural development, wars, and industrial transformation (Edmonds 1994). The loss of forest cover over the centuries led to severe soil erosion; in the early period of Chinese history, the Hwang Ho was not referred to as the Yellow River but simply as "the River" (Elvin 1993). Heavy erosion led to the addition of the adjective at the end of the first millennium B.C.

Timber was cut to meet the needs of the urbanizing cities in the northwest, using the forests found in southeastern China. By the period of the Tang Dynasty, population increases resulted in pressure to increase land arabilization, resulting in further deforestation. Deforestation caused soil erosion and, during certain periods, flooding and levees breaking.

Deforestation throughout history has caused topsoil loss and siltation, and when forest cover is depleted flooding usually follows (Roberts 1989; Ponting 1991).[9] The limited statistics for early China regarding population growth and recorded number of floods in the provinces reflect a high linear correlation (0.949) between population increases and the number of floods (see table 6.1). Regression analysis also shows that population increase is a good predictor of an increasing number of floods. The level of significance (r-square) is more than 90 percent. Placed within the wider context of the expansion and stagnation of

Table 6.1
Population of China and Number of Floods by Year A.D. 1–A.D. 1900

Year	Population (Millions)	Number of Floods
1	53.0	0
100	58.0	2
200	63.0	23
300	58.0	17
400	53.0	4
500	51.5	7
600	50.0	2
700	50.0	13
800	50.0	26
900	64.0	20
1000	66.0	42
1100	105.0	44
1200	115.0	72
1300	96.0	33
1400	81.0	81
1500	110.0	26
1600	160.0	52
1700	160.0	160
1800	330.0	172
1900	475.0	362

(Source: Chu 1926; Feuerwerker 1990)

growth over the long term, it is suggestive that environmental degradation also exhibits long swings correlating to population growth and economic expansion. The phases of economic growth in China from 1 A.D. to the present reflect a succession of logistic growth correlating with population increases. The limited data suggest that there have been four periods of surges (logistics) in population growth: from 400 B.C. to A.D. 200; A.D. 400 to A.D. 1200; A.D. 1550 to A.D. 1600; and A.D. 1700 or 1800 to the present. These population logistics are also repeated for Europe over the same period.

Within these long swings of population growth, the number of floods over time also exhibits increases and decreases over two hundred to three hundred years in length. Figure 6.1 and table 6.2 outline the tempo of these long swings over the last two millennia, with the number of floods increasing and decreasing over A/B phases of the system up to A.D. 1700[10] The limited data suggest that the number of floods increases during A phases and slows down during B phases. As the data are quite limited and restricted to one country, any further generalization would require further comparison over time and different geographic spaces.

For southern China, deforestation was the order of the day during the Ming and Ching Dynasties (Marks 1996, 1998). The various waves of migration and

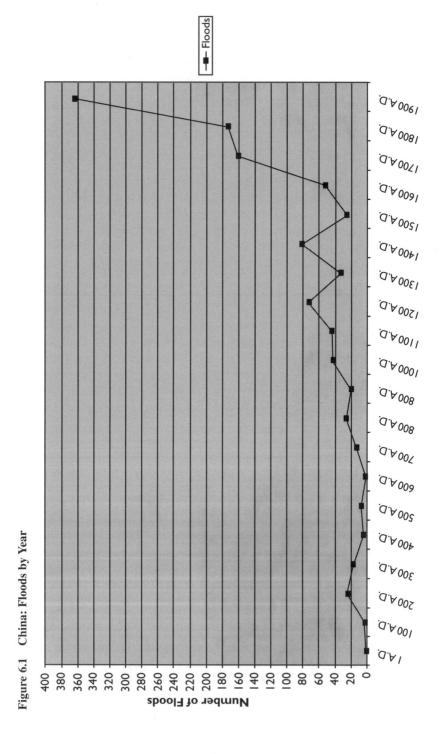

Figure 6.1 China: Floods by Year

Table 6.2   Number of Floods via A/B Phases for China 0–A.D. 1700

0–200	A phase	Increase
200–400/500	B phase	Decrease
400/500–750/800	A phase	Increase
750/800–950	B phase	Decrease
950–1200	A phase	Increase
1200–1500	B phase	Decrease
1500–1700	A phase	Increase

population increases from A.D. 742 to the seventeenth century A.D., and especially during the latter period, further exacerbated the toll on the landscape. The rate was so high that by the early nineteenth century wood was no longer available as fuel in parts of southern China (Marks 1996).

Besides the deforestation caused by agricultural cultivation and urbanization needs, manufacturing processes such as the iron industry also generated deforestation. Even as early as the Tang dynasty, just as in the earlier Roman empire, the iron industries were causing severe deforestation. By the middle of the Tang Dynasty, to meet their fuel needs, the iron manufacturing centers in Shantung, although they also used coal as a substitute, had to be moved to other areas where there were forests left (Hartwell 1967, 109). By the eleventh century, fuel sources continued to determine the location of forges and furnaces.

Deforestation continued with vigor during the Ming Dynasty, especially with the attack on forests in the north and northwest, which followed the move of the capital to Peking in A.D. 1420 (Menzies 1992). Besides deforestation as a consequence of agriculture, building and the need to reproduce exuberant lifestyles also evoked severe deforestation as early as the eighth century A.D. (Schafer 1963). Schafer (1963, 1967) has noted that the elites consumed exotic wood products — not only native species but also those from Southeast Asia. Later in the period, the construction of the Imperial Palace in Peking, which occurred between 1555 and 1600, required twenty-one thousand nan logs of varying length. The standard log size was about 82 feet in length and 13.5 feet in circumference (Winters 1974). These trees were removed from the province of Hunan and floated down the Yangtze River and then shipped north along the coast to the Peh Ho River.

Forest Conservation in Early China

Despite the severe deforestation in early China, there occurred moments in its early history of attempts to conserve the forest. As early as the Chow Dynasty (1122–256 B.C.), there were several initiatives to conserve the forests. Silvicultural practices were put into action, such as removing dead and defective trees, thinning dense groves, and eliminating undesirable species (Teng 1927). The Commission of Mountain Forests was established to institute rules and regulations to protect the forests and to prevent theft of wood. Enforcement was established by the formation of the Police of the Forest Foothills. Included with the foundation of these institutions was the passage of several imperial decrees prohibiting the cutting of timber and the starting of forest fires during certain seasons. Tree-culling practices were also enforced to ensure that tree maturity was safeguarded.

Such conservation measures were dropped during the Chin dynasty, primarily because of the extensive shift towards urbanization, agricultural production, and the building of public monuments and buildings. Later records indicate that conservation policies existed during the period of the Tang Dynasty (Schafer 1962). Parks were protected, including the animals living within such confines. Edicts were passed to prevent soil erosion from mountain slopes by prohibiting wood-cutting for fuel. Royal preserves were established to protect watersheds. In addition, reforestation was encouraged through policies to replace depleted groves.

Japan

The pattern of deforestation was repeated in Japan. Forests were cleared from 200 A.D. onwards for fuel (charcoal) to produce metal implements, for domestic use, for the cultivation of crops, and for building fortresses, temples, and shrines. To Totman (1989, 10), "agriculture and metallurgy were the human innovations that most dramatically affected prehistoric Japanese forests." Forest exploitation was continuous from A.D. 600 to A.D. 1670, and again in the first half of the twentieth century. As in China, deforestation was associated with economic expansionary phases. In Japan's case, there were three periods of cyclic economic growth and expansion (A.D. 600–850; A.D. 1570–1670; A.D. 1900–1950) that witnessed intense deforestation.

Massive deforestation started from the seventh century A.D. with the clearing of land for agriculture. The island of Honshu was severely stressed, especially the valleys, where agricultural production was being undertaken systematically. From A.D. 600 to A.D. 850 and onwards, within an economic expansionary environment, the ruling elites of Japan proceeded with a building boom, constructing palaces, mansions, shrines, temples, and monasteries. These buildings were erected near the capital cities of Nara and Heian (Totman 1989, 1992). By A.D. 628, forty-six Buddhist monasteries were built. It has been estimated that one hundred thousand *koku* of processed lumber were used to build a single monastery (Totman 1989, 17). Tokoro, as cited by Totman (1989, 17), has estimated that the three centuries of monastery building that started in A.D. 600 consumed ten million *koku* of processed lumber.[11] This interest began to ebb after a few centuries; interest in preserving woodlands by monasteries and local governments provoked by the massive deforestation also waned.

Continuous economic development led to the export of timber from Japan to China, which by A.D. 1000 had started to import wood to meet its domestic needs. By A.D. 1000 the population of Japan had reached 6.5 million, and it was to double by A.D. 1600 Such surges meant that wood resources were required to meet the growing population. The capital city of Heian faded away by the twelfth century after its surrounding forests were cut (Totman 1992, 19–20). To meet the socioeconomic expansion, Japan's old-growth forests were devoured by the seventeenth century. Monumental construction and urbanization continued in cities like Kyoto, Osaka, and Edo, whose populations ranged from four hundred thousand to one million. By 1720, the country's population increased to thirty-one million. Such population increases added further pressure on the forests so that by the end of the nineteenth century, with the exception of the island of Hokkaido, there was very little virgin timber left in Japan.

Conditions in China were repeated in Japan: erosion, extreme river silting, and flooding. Between A.D. 600 and 850, the mountains adjoining the Kinai basin were deforested, giving rise to wildfire, flooding, and erosion. Deforestation was so extensive that the emperor by 675 A.D. announced forest-closure policies to protect the remaining stands of trees. This forest-protection policy continued well into the ninth century, though overcutting, with its associated ecological consequences, continued until 1678.

As the forests were being removed in Asia, the same occurrences were being repeated in other parts of the world economy. The emerging economies of Europe also resulted in extensive deforestation. Such a pace began again in the fifteenth century, though as we have noted in the previous chapters deforestation of the European landscape had a long history. In addition to the increasing deforestation in Europe, resulting from the needs of the emerging European economies, we also witness at this time the "voyages of discovery" undertaken by the European states to the new world and beyond. Such ventures, especially as they resulted in forest exploitation, started in Asia with the arrival of the Portuguese in India.

NOTES

1. Abu-Lughod (1989, 251–53) sees it as an interplay of three interlocking circuits (stretching from the Arabian peninsula to southern China), each "within the shared 'control' of a set of political and economic actors who were largely, although certainly not exclusively, in charge of exchanges with adjacent zones."

2. Bayard (1992) has suggested that these developments might not be diffusion of technologies and ideas from China or India but rather a result of the cultural process and internal evolution.

3. Wheatley (1961, 17) has indicated that this state was peopled by families belonging to the Mongol and Tartar tribes of central Asia.

4. There is a question about whether this kingdom existed where it is supposed to have been located (see Abu-Lughod 1989, 315).

5. Abu-Lughod (1989) has viewed these tribute missions as "public" trading.

6. Wang (1958, 124–25) has indicated that Persian merchants had contacts with China as early as A.D. 518, but this was by land through the silk route.

7. The state was so supportive that it rewarded yearly renewal of trading licenses to Chinese merchants who returned with cargoes dutiable at more than five thousand strings of cash. Hotel accommodations were provided to foreign traders and welfare benefits to foreign shipwrecked seamen (Wheatley 1959, 25).

8. There have been a number of conjectures on why China withdrew from this area; see Abu-Lughod (1989). Finlay (1992) attributes the withdrawal to the internal conflicts between the civil bureaucracy and the eunuchs of the imperial court. To seal the fate of these voyages, the Confucian bureaucrats passed laws to make it a crime to build ocean-going ships with more than two masts.

9. It should be noted that floods are not caused only by deforestation; other factors engendering flood conditions, such as temperature changes and climatological shifts, can also account for floods.

10. The slicing of A/B phases over world history is derived from Frank (1993).

11. In terms of volume of wood utilization, one hundred thousand *koku* of processed lumber is sufficient to build three thousand ordinary 1950s Japanese-style homes (Totman 1989).

Chapter Seven

The Emerging Economies: Venice, Spain, Portugal, Holland, France, and England A.D. 500–A.D. 1800

The slow demise of the Roman empire took several centuries, and toward the end the northern portions of the empire came under severe border incursions from Germanic tribes (Anderson 1974; Braudel 1984). The first invasive wave started as early as the third century A.D., and these went on for more than two centuries. Strained ecological conditions caused by centuries of resource maximization and utilization coupled with social disruptions precipitated by external invasions led toward the end to a fragmented empire divided systematically among the Gauls, the Vandals, and the Visigoths. The breakup led to decentralized economies, whereas in the past it had been organized within a level of supply and demand along Roman lines.

Western Europe suffered through what is known as a period of darkness and emerged only following the attempt by Charlemagne in A.D. 800 to form a coherent economic system. At this stage, Europe was considered a backwater in comparison to other civilizations and empires in existence then, such as those in China and India. From the fall of the Roman empire until the thirteenth century A.D., Europe consisted of emerging economies participating in a world economy that had its poles in Asia.

Europe's emergence from a set of decentralized economies led also to the rise of small trading centers, which developed along the coasts to join those that dotted the trading routes. Such centers led to the reurbanization process, which had been stymied as a consequence of the Dark Ages and of Europe's prior decentralized state. The number of towns multiplied (Hohenberg and Lees 1985). Germany, according to Braudel (1984, 93), during this time had more than three thousand towns. These European towns further consolidated themselves with their existing Roman roads, markets, and manufacturing workshops.

With regard to trade, northern Europe was connected to the south by market-linking mechanisms such as the Champagne fairs, in addition to the trade conducted via the river systems that had been in existence since pre-Roman times

(Abu-Lughod 1989; Braudel 1984). Later on, northern Europe was linked by sea to the south via the ports of Bruges, Antwerp, and London. Bruges saw the arrival of Genoese ships by the end of the thirteenth century and Venetian galleys by the early parts of the fourteenth century. Along these firmly established land and sea links flowed the goods such as spices and pepper imported from the Levant and beyond by Italian merchants, who in turn traded for the industrial products of the north, such as textiles, grain, and wines.

Further south in Italy, Venice served as a crucial link between Europe and the expanding Moslem world (Braudel 1984). Venice became the trading gateway for products from the East and south. As others such as McNeill (1974) and Abu-Lughod (1989) have brought to our attention, Venice served as the crucial link between Europe and the East, especially after the fall of the Roman empire. Genoa, located on the western part of Italy, was a competitor to the Mediterranean trade for the products from the Levant and beyond. Its western location on the Italian boot provided it with a natural advantage for the western Mediterranean trade, especially with the Iberian peninsula. Venice, however, benefited from its alliance with the Byzantine empire and took full advantage of the commercial opportunities that lay eastward. It offered salt, fish, and timber from its own and European hinterlands in exchange for Eastern products.

Placed within such a context, the emerging economies of Europe managed to link themselves to the world economy, which had its centers in Asia as the preceding chapter has outlined. The series of Crusades embarked upon to subjugate the Moslems helped to enhance the Venetian, Genoese, and Pisan positions in the eastern Mediterranean trade starting from the thirteenth century onward. This trading exchange required the development of the exploitation of the natural resources and metals with which Europe was endowed at this stage of global transformation. Emerging economies tend to expand and export their natural resources and agricultural products, as their nascent industrialization processes are not yet producing market-competitive goods that the core accumulation centers are seeking. In addition, linkage into the world economy also engenders the need for fuel and other materials for shipbuilding and urbanizing. Again, such strategies often lead to deforestation. This is an issue we will examine in this chapter for the same time period as we did in the preceding chapter, but for a different geographic space.

ACCUMULATION AND DEFORESTATION

Venice: Shipbuilding and Deforestation

By the end of the fourteenth century A.D., Venice was becoming a center of commerce, at least for the Mediterranean, and assuming the mastery of the seas with its superior-quality ships. For Braudel (1984, 124) "the world-economy centred on Venice." All trade within the eastern Mediterranean and from Europe to the East went through Venice. The city was described as a warehouse of the world. The series of Crusades mounted to the Holy Lands enhanced the trade of Venice. The conquest of Constantinople by the Fourth Crusade in 1204 furthered the Venetian position when Venice claimed parts of the city and the island of Crete. Control of Constantinople and Crete meant that Venice was now in a position to control the lucrative trade with the East. From Crete, Venice directed her spice trade, which it controlled for the next two centuries. At this stage, the Venetian trading world stretched from the eastern Mediterranean to the Adriatic and from the Alps into northern Europe. To the south, its trading hegemony covered Egypt and eastwards; it controlled the lucrative access to spices and products from the East. By the thirteenth century, Venice had overshadowed Genoa and Pisa in the eastern Mediterranean.

Trade with Egypt and the East therefore expanded. Spices and luxuries from the East were sold by the Venetians to the wealthy urban classes of Italy and Europe, and in return the Venetians exported European timber, metals, and other coarse goods in which the East was deficient. In short, the Venetians became the middlemen of the East–West trade. Trading increases led to several transformations, such as urbanization and manufacturing. In the case of Venice, besides these economic endeavors, shipbuilding was also expanded.

Venice by the thirteenth century already had a history of shipbuilding, primarily in support of the Crusades. The increase in trade furthered this building process. In the early parts of the twelfth century, the Venetians constructed a municipal facility for shipbuilding: the Arsenal (Perlin 1989; McNeill 1974). It was a shipyard primarily devoted in its early history to manufacturing warships. Under state regulations, the Arsenal perfected the art of constructing ships with unusual skill and efficiency (Lane 1934). Workers specialized in one part of the construction process, and standardized parts were utilized for each ship being built, following the rib-and-plank type of construction. The process enabled quick repair and construction; eventually each ship could be constructed in less than an hour (McNeill 1974, 6).

By 1300, with technological innovations in ship construction, northern Europe was further connected to the Mediterranean, and so was the Black Sea region. The stern-post rudder was one such feature. Such innovations were also facilitated by new navigational techniques, such as the adoption of the compass, which enabled directional finding under cloudy skies (McNeill 1974). This led to year-round shipping and the fostering of a carrying trade.

The integration of northern Europe with the Mediterranean added further push to economic transformations throughout the region. The expansion of the wine industry and textiles manufacturing in northern Europe further accelerated the economic dominance of Venice and its supremacy between 1282 and 1481. All such economic activities, along with slave trading, naturally required transportation via sea routes and land portages.

Because the ships were built in the Venetian state-sponsored shipyard, the merchant galleys were owned publicly. This being the case, the hauling of freight such as spices and luxurious items was restricted to only state-owned ships. Therefore, a tight control was exerted over the profits, which were quite substantial. According to Doge Tomasso Mocenigo's statements, in 1423 on ten million ducats invested annually a profit was made of more than four million ducats (McNeill 1974, 61). Besides engaging in merchant shipping, Venice, trying to maintain its political hegemony, also had a standing army during war times and a large navy. During certain periods, such as the war of succession to the visconti of Milan in A.D. 1450, the navy amounted to more than one hundred galleys.

From the thirteenth to the fifteenth centuries A.D., Venice was the major power in the Mediterranean and Europe. Its main industrial activities lay in shipbuilding, glassware, soap, silks, and jewelry. It also acted as an entrepôt for textiles and woolen cloths from northern Europe and metals from Germany. Venetian merchants also invested in agricultural plantations that grew sugarcane on the island of Cyprus and cotton in Syria, Anatolia, and the southeastern Balkans.

The Arsenal continued to build merchant galleys and warships to sustain Venice's power. To satisfy the wood resources of this enterprise, the surrounding forests provided the wood and pitch. The Venetian glass industry also required an extensive amount of wood, which led to deforestation of the nearby forests. To feed the growing population, pastureland had to be cleared. Venice's population had grown tremendously. From about 80,000 persons in 1200 it doubled to 160,000 persons by 1300. This growth slowed tremendously when Venice lost its status as a major economic power by the fifteenth century, when its population hovered around 160,000. It lost about 45,000 persons to the plague in 1575 and grew again to about 180,000 by the end of the sixteenth century.

The contest between Venice and the Ottoman empire also engendered heavy deforestation to meet the shipbuilding needs of Venice. From 1470 to 1492, a policy of conservation was introduced on state woodland to ensure that a supply of oak was available to meet shipbuilding needs. Such acts failed to prevent Venice's decline, and after 1479 it sunk to secondary rank in terms of power and prestige in the Mediterranean. By 1481, the Turks had defeated Venetian strength on land and sea. The slow slide by Venice was a consequence of a series of factors, such as the penetration of Asia by the Portuguese in Calicut between 1497 and 1499 led by Vasco da Gama. These voyages led to certain upsets in the spice trade, over which in the past the Venetians had held a monopoly. However, by 1502, the flow of spices and other condiments to the Mediterranean was threatened by Portugal's incursion into the Indian Ocean and its attempt to dominate the spice trade and the riches of south Asia and beyond.

With control of the spice trade in a precarious situation, the Venetian slide continued in the area of the merchant-marine and carrying trades. Venice's loss of dominance in the carrying trade commenced around 1590, when British and Dutch ships started entering the Mediterranean in large numbers, conveying produce from northern Europe (McNeill 1974; Braudel 1972). The loss of Venetian dominance in the carrying trade was due mainly to the scarcity of wood encountered by Venetian shipbuilders starting around 1530, when they had to pay twice the price for wood as their predecessors (Perlin 1989; McNeill 1974). After the mid-sixteenth century, Venice found that its surrounding areas were depleted of wood, so it had to rely on timber supplies from Istia, Dalmatia, and other areas of the Adriatic coast. As a consequence, the pace of shipbuilding had to be scaled back, which affected the accumulation processes. Furthermore, later in the century, Venice responded to this competition by allowing Venetian ships to be built in northern Europe, especially in Holland, thus giving the Dutch further opportunities for their ascent. This they accomplished when Holland became a center of accumulation later during this period. Witnessing the loss of Venetian supremacy, the Venetian state also tinkered with various tariffs and taxes to make Venetian ships competitive in the carrying trade. These efforts proved fruitless; by 1610 Dutch and English ships had seized the majority share of the merchant-marine shipping trade.

These conditions continued to plague the Venetian economy until the seventeenth century, when the Dutch charged into Southeast Asia. Under the aegis of the Dutch East India Company, the Dutch cut off the spice trade much more effectively than the Portuguese had earlier. The Dutch ascent was primarily due to the efficiency of its shipbuilding process and its access to wood resources. With its easy reach of the forests of northwestern Europe and the Baltic countries and its efficiency in construction costs, Dutch ships could convey spices around the Cape of Good Hope more efficiently than the Venetians, who relied on traditional routes that proved limited.

To Perlin (1989), the continued lack of access to wood further hampered Venice's position in the overall accumulation process, and was one of the factors that led to its decline as a major trading center (center of accumulation). McNeill (1974) also regards such conditions as being among the factors affecting Venice's ability to maintain its position within the world economy at that period in time. Thus, centuries of overexploitation of surrounding forests resulted in an ecological limit, leading to shortages of timber and wood for shipbuilding and for fuel. The outcome impacted on the population and industries of Venice. Venetian decline was completed between 1769 and 1797. The commercial power vacuum left by Venice was filled by northwestern Europe, a phenomenon that also shifted one of the geographic loci of the accumulation process in the world economy away from the Mediterranean. By this time, globalization linkages had been solidified further, with maritime trading routes established to the Americas and Asia from northwestern Europe by Portugal, Spain, Holland, France, and England (Frank 1998; Modelski and Thompson 1996).

NEWLY INDUSTRIALIZING ECONOMIES
OF NORTHWESTERN EUROPE

Europe in Asia

Prior to the final decline of Venice in the late eighteenth century, the emerging economies of Portugal, Spain, Holland, France, and England were starting to develop economically and transform themselves. By the fourteenth century, the forests of southern England had regenerated from the depletion they sustained for use in the Roman smelting works of centuries before, and England was exporting wood to Holland and France. The growing rise of Holland as a center of accumulation meant that England was part of the peripheral hinterland, supplying wood to fuel Holland's needs. English wood, especially its oak, was in high demand, and six hundred shiploads left England for Holland each year. England at the time had few industries requiring wood, as it exported natural resources and imported finished goods (necessities and luxuries) from Europe.

However, later on in the period, English consumption of wood accelerated with the development of munitions production in southern England, around Sussex, under the reign of Henry VIII. With the production of iron for the manufacture of cannons, wood consumption shot up, for a ton of bar iron required forty-eight cords of wood. Added to this consumption was that which occurred to support the increased production of copper, salt, and glass, and of ships for the Royal Navy and the merchant marine. A large warship required two thousand oak trees at least a century old (Chew 1992, 25). Besides utilizing its own wood resources, England on its ascent as an economic power also relied on and exploited the forests of Ireland.

Portugal by the end of the fourteenth century was also on the ascent in terms of economic growth. Its incursion into Africa, with its profitable resources of ivory, malaguetta (a pepper substitute), gold, and human slaves, generated economic rewards that furthered Portugal's explorations of Madeira and the Azores in the early part of the fifteenth century. As we indicated in this chapter's previous section, the discovery later of a sea route to Asia via the Cape of Good Hope furthered the rise of Portugal in northwestern Europe.

The development of sugarcane plantations in Madeira by the Portuguese is an exemplar of island deforestation as a consequence of accumulation of capital by a colonial power. When the Portuguese arrived at Madeira, the island was so thickly filled with trees that the Portuguese named it "island of timber" (Perlin 1989, 249). Trees were removed to develop sugarcane plantations. The extensive deforestation was furthered when wood was required to produce sugar; wood was required in construction of the sugar mills, the machinery that squeezed the cane, and the waterwheels that powered the machinery. In addition, a large amount of energy was required to boil the juice to transform it to solidified sugar. Such an array of requirements necessitated the need for large amounts of wood, and by the end of the fifteenth century about sixty thousand tons of wood were required per year to produce sugar in the island's sixteen mills. In addition, wood was exported to Portugal for the building of larger ships. This facilitated Portugal's mastery of the seas and its voyages of exploration and enabled it to enter the spice trade of Asia. The move into Asia challenged Venice's monopoly of the trade and reduced its control. Portuguese penetration of Asia in 1498—when they landed first at Calicut, then subsequently at Goa, Malacca on the Malayan Archipelago, and Macao—led to the cutting of trees to meet the needs of the Portuguese fleet (Gadgil 1988, 49). The shortage of wood in southern Europe stimulated Portugal into building ships in India, utilizing the abundant quantities of Indian teak (Boomgaard 1998).

When Portugal entered the trading world of Asia it needed to establish a foothold in a centuries-old commercial network conducted by Asian merchants from different ethnic origins: ranging from Armenians, Bengalis, and Gujeratis to Javanese and Chinese (Curtin 1984; Chaudhuri 1985). Portuguese interests in Asia were controlled and administered by the establishment of the Casa da India and the Estado da India. The former was responsible for commercial operations while the latter was in charge of political and military activities, including the administration of Portuguese forts and settlements. Though not totally successful, the Portuguese captured a portion of the overall Asian spice trade to Europe (with the exception of pepper); by the 1560s about half of this trade was handled by the Portuguese (Curtin 1984). In the case of pepper, by the latter part of the sixteenth century the Portuguese controlled over three-quarters of the commerce to Europe.

Around 1521, Spain also had an interest in Asia, centered mostly on the Philippines. Spanish interest was an offshoot of its operations in the Americas followed by Magellan's arrival in the islands. Like the Portuguese, the Spanish crown was also keen on getting a slice of the spice trade of Asia. Spanish interests thus competed with the Portuguese. Losing this competition, Spain later focused its efforts on the already existing trade between the Philippines and China; it used gold and silver from its Mexican operations to trade for Chinese silk, porcelain, cotton cloth, and other wares. Thus rose the famous galleon trade that continued until the early part of the nineteenth century. Spanish colonization of the Philippines led to severe transformation of the natural landscape. Besides building construction, the utilization of wood for shipbuilding saw the island of Cebu stripped of its hardwood to build Spanish galleons (Tadem 1990, 15).

The Portuguese and Spanish expansions also detonated the emergence of Holland as a major player in northwestern Europe. As indicated earlier, the Dutch built many ships for Venice when Venice faced a wood shortage. The extent of their ship construction led Holland to be dubbed as the "the Arsenal of Europe" (Perlin 1989, 160). By the end of the seventeenth century the total size of the Dutch fleet amounted to about six thousand ships manned by forty-eight thousand sailors (Braudel 1984, 190). To Braudel (1984), this fleet became the instrument of Holland's ascent in its overseas expansion. Dutch ships were more economical to build and operate as they needed fewer sailors than, for example, French ships, which in most cases required at least a third more sailors for the same tonnage of ships.

Starting in 1595 and initiated first by merchants, Holland's entry into the Asian trade was quite unorganized. Merchant ships were sent to tap into the rich spice trade that the Portuguese had captured. These trading excursions were consolidated with the founding in 1602 of the Dutch East India Company (Vereenigde Oost-Indische Compagnie or VOC) (Braudel 1982; Chaudhuri 1978). Even so, Dutch expansionary trade into Asia had to contend with existing Portuguese forts and settlements already established in Asia. As the Portuguese had, the Dutch were trying to penetrate an established trading arrangement conducted by Asian merchants, from Turks and Armenians to Arabs, Bengalis, Chinese, and Javanese.

The 1596 Dutch reconnaissance voyages that reached the port of Batavia in Java were pursued further in 1619 by the Dutch East India Company. The Dutch colonization of Java in the seventeenth century led to the cutting of teakwood to meet initial requirements of the Dutch East India Company, and later the imperial needs of Holland (Peluso 1992a). Wood was sought in the initial stages for shipbuilding, building construction, furniture making, and for charcoal fuel. Further consumption is revealed by the construction of shipyards; by 1675 we find one employing more than 250 persons and the presence of wind-powered sawmills. In concert with local rulers and middlemen traders, the Dutch East India Company harvested the teak trees, which occupied about 6 percent of the total surface area of Java. Other trees, which covered approximately 17 percent of Java, were deemed worthless junglewood and considered suitable for destruction, "of which nobody suffers" (Boomgaard 1988, 61). When the Dutch East India Company ceased to exist in 1800, the control of the forests of Java was turned over to the Dutch Republic.

Besides Javanese products, the Dutch East India Company's operations in the Dutch-Asiatic trade centered around bullion, wood, pepper, spices, tea, coffee, copper, silver, silk, and raw cotton (Glamann 1958). To produce most of these products required the transformation of the natural landscape. Forest products were among the main items that formed the mercantile activities of the Dutch East India Company. Boomgaard (1998) has estimated that for selected three-year periods between 1619 and 1780, these products represented 1.8–9.8 percent of the total value of cargoes onboard VOC ships bound for Europe. A century later, the volume increased to 5.8–9.8 per cent. The increase meant that overall the trade in forest products was more important than that of other commodities, such as sugar and metals (Boomgaard 1998).

In addition to exporting Asian products to Europe, the VOC also participated in intra-Asian trade. Brazilwood, which was abundant in Siam, was also shipped to Japan for consumption by the VOC. Between 1633 and 1663, 160 tons per annum were shipped to Japan. In the later part of the seventeenth century, this amount was reduced to about 65 tons annually (Boomgaard 1998, 380). The volume of brazilwood in the intra-Asian trade was dwarfed by what the VOC exported from Siam to Europe. Over the same period, 400 tons were shipped annually to Europe until 1663, when the annual load was reduced to about 250 tons till the end of that century. Bima was also a source of brazilwood for the European markets. From there, an annual volume of 800 tons was shipped out during the same period until 1663. Other types of wood were also exported from Southeast Asia by the VOC. Ebony and sandalwood were sent to Europe and other parts of Asia such as Japan, China, and India, where they were sought for ceremonial and religious purposes. Exports of ebony from India to Holland amounted to about 250 tons per annum during the latter half of the seventeenth century, while from Batavia during the same period sandalwood from Timor, Makassar, and Java was sent to other parts of Asia such as Japan and China. The volume of sandalwood exported reached about 100 tons annually.

Payments for forest products were usually settled in gold or silver. Besides silver from the Americas, gold and silver were also obtained from local sources. Dutch traders also relied on Japanese silver and gold for their transactions. Sumatra and the Malayan archipelago were also supply sources (Braudel 1984, 217). In this light, it was not only silver from the Americas that was utilized to finance the European trade in Asia, as Frank (1998) and Blaut (1992, 1993) have asserted repeatedly; the Dutch relied on gold and silver from local indigenous sources—such as the Malayan archipelago, the Indonesian islands, and Japan.

Paralleling the Dutch entry into Asia was England's. The English East India Company was formed in 1600, two years earlier than the VOC. Undercapitalized initially, the English East India Company at first was not competitive with Dutch operations. English operations were centered around Surat in India and the islands of Ambon, Makasser, and Java in the Indonesian archipelago. Later it expanded to Bombay, Calcutta, and Madras. Factories were established for the spice trade in Indonesia. English presence in Southeast Asia, in competition with the Dutch, often turned violent. The massacre of the English at Ambon by the Dutch led to an English retreat to Banten in 1628. Such setbacks indicate that the English East India Company was never a challenge to the VOC. By 1682, the company had even abandoned its operations in Banten and retreated to India, which later became its principal commercial area.

The European entry into the trading world of Asia exacerbated the transformation of the natural landscapes. By no means can it be assumed that the landscapes of Asia were in a pristine form prior to the arrival of the Europeans. In the previous chapter, we have outlined the scale and breadth of economic activities undertaken by the Asians prior to the arrival of the Europeans. In fact, Asian merchants from the Indian Ocean to the South China Sea continued to be extremely active despite the arrival of the Europeans. The expansion of the Chinese economy from the Tang Dynasty to the Sung and Ming periods primed Asian trade. Whole communities in Asia were devoted to producing commodities for the export market, from wood products to pepper, cotton, sugar, etc. The growing of cotton was widespread. It was grown in the Indonesian archipelago and also in the Philippines. From these areas, raw cotton and cotton cloth were exchanged all over Asia, even in China, whose textile industry was very developed by this time. Other cash crops, such as sugarcane, were grown extensively in most Southeast Asian countries and then exported to Japan and China. Thus, the arrival of the Europeans further intensified an already ongoing degradation of the land and forests. From the vantage point of world history, such a scale of economic activities of the Asians and the Europeans caused deforestation of the natural landscapes of Asia, and especially of Southeast Asia, where most of these products were sourced. Deforestation as a consequence of accumulation was by no means a new feature in Asia for this period in history.

Europe in the Americas

In addition to Asia, the rise of the economies of northwestern Europe was also facilitated by the natural and social resources of the New World. Prior to Vasco da Gama's rounding of the Cape of Good Hope, initiating the entry of the newly emerging economies of northwestern Europe into the trading world of Asia, Columbus's arrival in North America in 1492 linked the Americas with Europe. From then on, long-term and large-scale transformations of the landscape by European powers occurred concurrently in several areas of the Americas, from North America through the islands of the Caribbean to South America.

Benefiting from Columbus's voyages, the Spanish explored and conquered the New World with enormous ferocity under such conquistadors as Cortez and Pizzaro. Setting up colonies from the coasts of Chile and Argentina to Ecuador, Colombia, Venezuela, Central America, and the southern United States, the Spanish proceeded to extract the natural resources and bounties of the landscape. Of course, the forests of these areas suffered under such a scale of economic activities. Such penetration led by the end of the sixteenth century to the voyages across the Pacific to the Philippines.

Included in this early colonization process were the Portuguese, who landed in Brazil and began to settle the country after 1530. In Brazil, a plantation economy was started with sugarcane as the basic cash crop along with tobacco. By 1709, twenty-five thousand tons of coffee were exported annually from Brazil. In addition to sugarcane and tobacco cultivation, cattle ranching was also pursued with much vigor. Ranching as a commercial activity requires extensive clearing of forested land. Approximately twelve million acres were removed (Dean 1983). The natural resources of Brazil came under assault, with the gold of Minas Gerais as a prime resource. Diamonds also encouraged the economic development of the colony. All of these economic activities, like those we have encountered in the preceding chapters in other parts of the world and at different periods, naturally engendered the deforestation of the landscape. It is estimated that nearly five million acres of forests were cleared for mining activities. Additionally, brazilwood, which produced a dye for textile manufacturing, was also sought after by the Portuguese. Following on the heels of the Iberian powers were the Dutch, the French, and the English. Holland, however, focused most of its efforts in Asia and had some interests in the Americas. It is the French and the English that penetrated the New World and maximized their economic advantages through the establishment of colonies in North America, the Caribbean islands, and some parts of South America. The colonization of mainland North America by the French in 1608, the English in 1607, and the Dutch in 1613 challenged the power of Spain and its monopoly in the New World.

The New World offered the French and the British bountiful natural resources in addition to vast expanses of land available for cultivation of cash crops (sugarcane, tobacco, indigo, and cotton). The islands of the Caribbean—such as Antigua, Nevis, Montserrat, Barbados, Jamaica, Dominica, Saint Vincent, and Grenada for the English, and Martinique and Guadeloupe for the French—provided the same bounty of natural resources for undertaking economic export activities. In New France, the Intendant promoted shipbuilding and sawmilling in line with French colonial policy of ensuring a positive balance of trade for the mother country (Ryerson 1975; Lower 1977). Large-scale operations were started as early as the 1650s, with the exportation of square beams and barrel staves. Shipbuilding was fostered; ships of 120 tons were built in Quebec. To ensure that the bountiful supply of trees in New France remained available, the Intendant was ordered to survey the forests for the French navy.

By the mid-seventeenth century, England's and France's indigenous wood supplies were running low after centuries of wood exploitation to meet export and industrializing needs (Chew 1992). The scarcity of indigenous supplies in England and France had been reached as a consequence of the previous century of growth—the "long sixteenth century" of expansion. In England's case, deforestation was "encouraged" further by the enclosure movements and the intensive growth in farming prior to the mid-seventeenth century, which led to an extensive cutting of trees (Chew 1992; Albion 1926). Furthermore, the planting of oak trees was not as profitable a venture as the growing of corn, since an investment in oak trees would not be realized until a century later. Wood was used extensively in the construction of buildings and in manufacturing. It was, as some would say, an "age of wood" (Lower 1973). Tanners, soap boilers, gunpowder manufacturers, and glassmakers all required wood products, such as oak bark and potash, for their operations. The mining industry needed sturdy timber for its mine shafts. Brick manufacturers, sugar refiners, and salt producers used wood. Fifty cubic feet of hardwood was required to produce two thousand bricks, and more than a load of hardwood was needed to produce two hundred weights of salt (Chew 1992, 23).

In 1669, France, suffering from a scarcity of indigenous supply, passed the Forest Ordinance, declaring a moratorium on cutting the forests throughout

France for eight years (Chew 1992; Grove 1995). Whereas the French dealt with their wood-scarcity crisis via better management of their forestlands, England adopted a different approach. Its policy was to displace its timber needs by relying on external wood resources in two continents: northeastern Europe and North America. As a result, with increasing pressure arising from internal scarcities of timber and lumber, the forests of these areas became targets for assault. Thus, the fir forests of the Baltic shores and the oak belt found to the south and west, along with the forests of New England and eastern North America, became important regions for British timber supplies. At this stage, the Baltic area was preferred because of its proximity and the lower shipping costs for transporting the wood products.[1]

In the Baltic region, the ports in Eastern Europe and Russia that were major timber exporting centers were Gothenburg, Oram, Christiania, St. Petersburg, Stettin, Danzig, Memel, and Riga (Chew 1992). The first four ports exported wood mainly from fir trees, while the rest sent a mixture of fir and oak. On the Scandinavian peninsula, spruce spars and fir timber were exported from the ports of Longsound, Dram, and Friedrikstad. The timber supply was facilitated further by trade agreements between England and Russia reached in 1734 and 1766. Through these agreements, the English were granted privileges identical to the ones enjoyed by the local merchants in Russia, which led the native Riga merchants to lose their monopoly of the mast trade. In fact, foreign buyers would commission English merchants to handle their purchases.

Supply of wood was also ensured by the English government's reserving and legislating protection and ownership of the forests of North America, where England had its colonial settlements. In 1705, the English Board of Trade and Plantations passed the Naval Stores Act, which provided bounties and preferences to stimulate the production of timber in the colonies of British America. Later, for example in 1722, the British government passed a legislative act prohibiting the clearance of white pine trees by settlers in its colony in British North America (Chew 1992). In 1763, after the conquest of New France, legislation and instructions were included in a dispatch to General Murray, military commander of New France, from the British Government, ordering him to reserve the woodlands for the Royal Navy, which by then was very dependent on foreign wood for its shipbuilding needs. By then, the Royal Navy tonnage had increased to 325,000 tons from 100,000 tons in 1685 (Hobsbawm 1968, 50).

Not only was wood needed by the mother country, English plantation owners in the Caribbean islands such as Barbados also needed wood for sugar-manufacturing operations. As in Madeira under the Portuguese, island deforestation for the growing of plantation crops was the order of the day for the Caribbean islands under British control. Besides local island supply, New England and eastern North America provided timber for manufacturing plants, staves for casks and barrels, and lumber for housing. Consumption of wood in the West Indies was quite exorbitant. More than 240,000 trees were cut in New England between 1771 and 1773 for consumption in the West Indies. Volume of wood products consumed in the West Indies for this time period amounted to seventy-seven million feet of boards and timber, sixty million shingles, and fifty-eight million staves.

Wood thus was an important commodity in the ongoing triangular trade between England, New England, and the Caribbean. New England's wood and English manufactured products were exchanged for the produce of the Caribbean such as sugar, rum, and molasses. In turn, New England received manufactured products from Britain in exchange for wood, sugar, rum, molasses, and fur. Placed within this context, the deforestation of the landscape was undertaken to meet imperial needs of Royal Navy ships, British commercial interests, and the growing building and manufacturing needs of Great Britain, the colonies in the Americas, and the Caribbean islands.

French and British colonization of the Americas and the Caribbean islands led to the transformation of the landscapes; the islands especially experienced widescale deforestation. This is a repeat of what occurred in Madeira under the Portuguese. Because of the wide array and expanse of forests in the Americas, deforestation occurred extensively to meet not only the varied needs of the colonies but also the industrializing demands of France and Great Britain in the seventeenth and eighteenth centuries and onward. This scale of transformation continued unabated in view of the extensive areas available for resource extraction. The subjugation of the native population via force and conversion to Christianity ensured the continuous extraction processes to meet colonial socioeconomic needs. The pacification process of the indigenous populations was facilitated further by the European transmission to them of diseases such as smallpox and measles (Crosby 1972, 1986; McNeill 1976).

An examination of these landscape-transformative patterns would lead one to signal the specificities of these epochs in terms of Nature–Culture relations if the examination were restricted exclusively to this period, without any reference to the macro-horizontal integration of socioeconomic processes on the world scale over world history. However, if consideration is given to locating these transformative epochs within the anthropocentric patterns of the long history of human communities' relations with Nature, such trends continue the tendencies that have occurred over the last five thousand years of world history. Therefore, in the following chapter, which examines the next two centuries of what is considered the "European Age," we will see that such trends persisted—except their trajectories followed an even higher level of degradation.

NOTES

1. For a detailed discussion of the political economy of timber and lumber production in this area, see Albion (1926) and Chew (1992).

Chapter Eight

Europe at the Helm: A.D. 1800–A.D. 2000

The world historical processes of accumulation, urbanization, and population growth spiraled upwards over the centuries, culminating in a much transformed ecological landscape by the beginning of the nineteenth century. By this time period, the patterns of trading and manufacturing relationships, political alignments, and cultural differences were increasingly organized by a small number of countries located in northwestern Europe, and joined in the twentieth century by the United States of America. In short, the world was shaped increasingly by what is commonly categorized as the West.

The global system that has developed over the millennia became even more connected than before, with its guidance shifting from the East to the West as the nineteenth century opened. Concomitant with this maturation, we find the same transformative outcomes upon the landscape, albeit at a more accelerated level than before, and at an even more interconnected level. Whereas in the past, ecological degradation occurred not only in the urbanized areas, but also in the peripheral areas where trade and imperial rule held sway, the geographic limits of ecological decay now became global due to the political-economic empires constructed by European powers that spanned the world at the dawn of the nineteenth century.

History thus continues to repeat itself in the area of the Nature–Culture relations, except that for this time span we witness an accelerated pace of landscape transformation that had not been seen before in world history—a consequence of the rapid trajectories of capital accumulation, population growth, and urbanization on a global, interconnected scale. Landscape degradation was expedited by technological advancements and innovations (mechanization and mass production) that occurred over the millennia from the Fertile Crescent to China and Europe. For example, forests could be cut down at a much more rapid pace utilizing mechanical equipment powered by motive force such as steam, and, later in the twentieth century, gasoline for chainsaws. Electricity replaced water as the energy source for running lumber mills. Changes at the turn of the nineteenth century, such as the utilization of electricity,

enabled mill operations to run around the clock, thus increasing the pace of production and also the velocity of realization of capital invested (Chew 1992). The utilization of steamships and railways for transportation enabled the shipping of forest
products on a year-round cycle, whereas in the past, when the motive force was the
wind, sailing ships had to wait for the spring thaw and the prevailing monsoon winds.
All in all, ecological degradation was intensified across the world system.

As we come to the end of this historical journey, we return full circle toward
understanding world ecological degradation by viewing again the prism of deforestation as a key proxy to analyzing the intensity of Nature–Culture relations.
Forest removal is a consequence of a number of human activities, as we have indicated in the previous chapters. For a system driven by the processes of accumulation, urbanization, and uncontrollable population growth, resource depletion
and degradation become again the predominant macroparasitic theme that defines
the Nature–Culture relations in world history, and especially for the time period
we can describe as the "European Age." Deforestation was global during this period due to the connectivity circumscribed by the political-economic relations established throughout the nineteenth century in the form of colonial rule by European powers such as Great Britain, France, Holland, Belgium, Portugal,
Germany, and Italy. The trend continues through the first half of the twentieth
century, when the United States of America can be included in this power matrix.

GLOBAL DEFORESTATION IN THE NINETEENTH CENTURY

The colonial adventures started by the West in the late fifteenth and early sixteenth centuries intensified and expanded in the nineteenth century. Trading patterns and colonial settlements were entrenched in the main regions of the world;
some colonies, such as in the Americas, achieved independence. Despite political
changes and alignments, capital accumulation continued unabated with the mil-
lenia-old practice of resource extraction in peripheral areas to support core consumption and reproduction as the norm; at this point in time, the extraction was
global in scale due to the world-encompassing nature of the colonial empires that
had been established. Depletion of resources such as timber and minerals was by
no means the only ecologically destructive activity undertaken. The clearing of
land for cash-crop plantations, whose exports depended on the needs of the colonial power, was a prime factor in the deforestation process. With accumulation,
we also witness the parallel processes of urbanization and population growth for
some and losses for others, which increased pressure to deforest to meet the expansionary reproductive needs of a system on an exponential growth trajectory.

It is very difficult to estimate the loss of forests on a global basis. Global population from 1700 to 1800 rose from 679 million to 957 million. From 1800 to
1900, it grew to 1.65 billion—an increase of around 72 percent (see table 8.1).
Notwithstanding the needs of the accumulation process, these increases led to

Table 8.1 Global Population Levels (Millions) A.D. 1700–2020

	1700	*1800*	*1900*	*2020**
Africa	107	107	133	1,441
Asia	435	630	925	4,680
Europe	92	145	294	514
USSR	30	49	127	343
North America	2	5	90	327
South America	10	19	75	719
Oceania	3	2	6	37
Total	679	957	1,650	8,061

* Projections
(Source: *The Earth as Transformed by Human Action,* edited by Billie Lee Turner, et al., table 3-1, p. 42. Copyright © 1990 by Cambridge University Press. Reprinted by permission.)

further pressures to meet reproductive needs, which in turn meant clearing forests for fuel and timber and for the planting of crops.

Rough estimates utilizing the distribution of land use can provide a sense of the overall loss. Williams (1997, 171), estimating deforestation levels, gives us a rough estimate of the amount of forest losses. Between 1700 and 1850, the area of crop land increased globally by 126 million hectares with a further increase of 230 million hectares by 1920. This increase in crop land was preceded by forest depletion of 128 million hectares between 1700 and 1850 followed with an even higher increase of 134 million by 1920. In total, forest loss over a 220-year period was 262 million hectares.

European expansion to secure political and economic resources impacted on the forests globally. On a world basis, nonnative species of crops, plants, trees, and animals were transferred from one region to another for cultivation to generate exports for human consumption and manufacturing needs. For example, sugarcane was introduced to Latin America and the Caribbean from Southeast Asia. Rubber trees were introduced to Southeast Asia from Brazil. Sheep, goats, and cattle from the Old World were transplanted to the settler colonies such as Australia and New Zealand (Clark 1949). All such activities resulted in forest loss from the clearing of land either for cultivation or for grazing. Furthermore, deforestation occurred to meet the energy, building, and transportation needs of the colonial powers either in the far reaches of the empires or within the mother countries. Thus, the impact was extensively felt throughout the world wherever European expansion occurred.

REGIONAL IMPACTS

Europe and North America

As seen in previous chapters, deforestation has been a long, continuous process in Europe, with some periods of forest rejuvenation due to decreases in political-economic activities during specific epochs, such as the Dark Ages. Thus, the forests of Europe had already been under assault for a long period of time. The growth in population from 71 million in 1600 to 145 million by 1800 led to extensive land clearing. Along with expanding industrial activities, ship-building and mining generated increased demands for raw materials and energy. What resulted was that by 1800 only about 5 to 10 percent of the European landmass, including Great Britain, remained forested. Such a lowered level of available forests forced the British to rely in some cases on forest resources beyond Europe to meet the reproductive needs of the imperial powers.

The nineteenth century furthered such reliance. The Napoleonic Wars, fought between Britain and France, led to a tremendous shift in wood sources for Great Britain. It had to switch its reliance from northeastern Europe, the Baltic countries, and Russia to North America (Chew 1992). This move, prompted by the Continental Blockade—issued by Napoleon in 1807—led to the shifting of production operations by British timber interests to North America; in the past they had operated in the Baltic countries and northeastern Europe and Russia. The consequence was a heightened level of cutting, especially in the forests of northeastern North America, such as in New York, Maine, Vermont, New Hampshire, New Brunswick, Nova Scotia, Quebec, and Ontario. In short, North American forests made up for the loss in wood supply resulting from the Continental Blockade. The Carolinas in the southern United States also participated as a source of wood.

The United States, as a settler colony, had in the past supplied wood to Great Britain and the Caribbean even before the nineteenth century. The nineteenth century saw wide-scale transformation in the United States as European and British immigrants started arriving as a consequence of the various wars and famines that ravaged Europe and Great Britain during the early part of the century. Land clearing was actively carried out to provide land for cultivation. By 1850, more than 46 million hectares were cleared. This land came from forested areas, only a small amount of which resulted from natural or Native American clearings. The pace continued unabated. Between 1850 and 1860 more than 16.1 million hectares were added to the total. The forests of New York, Ohio, Indiana, Illinois, and Wisconsin came under the axe. Total deforestation for the whole century amounted to more than 120 million hectares. According to Williams (1997, 173), "it was one of the greatest deforestation episodes that the world has ever seen." Besides land clearing for crop cultivation, forests were also removed for energy needs, such as heating, and for industrial uses. Wood in the form of charcoal was needed in large quantities for iron furnaces. By 1856, of the 560 iron furnaces operating in the northeast, 439 were utilizing charcoal as fuel (Williams 1990).

The pace of iron production continued throughout the nineteenth century; this, coupled with the operations of indigenous timber and lumber companies, meant that the assault on the forests of the eastern seaboard and in the Midwest of North America continued until the early part of the twentieth century. Deforestation occurred, though these areas had lower population densities and thus were less impacted in terms of floods and other consequences. After the turn of the twentieth century, the exploitation of the forests of North America progressively moved to the Midwest and the West Coast. Wood exports of primarily Douglas firs and coastal redwoods were shipped from the Pacific Northwest to Asia and beyond.

Latin America: The Case of Brazil

The deforestation of Brazil was already ongoing as the nineteenth century commenced (Williams 1990; Dean 1983; James 1932). The country that occupies the largest landmass in Latin America provides the overall tone for understanding the dynamics and character of deforestation in Latin America during the nineteenth century. By the 1830s, coffee, which was introduced earlier from East Africa, had supplanted sugarcane as the main cash crop. The coastal forests of Minas Gerais and Espirito Santo were the first target areas to be cleared of forests and replaced with coffee plantations. Forests located on hillsides were preferred choices because of the cooler air, but this caused devastating soil erosion, which led to abandonment of the land. With a large expanse of land available, forest removal proceeded without any restraint, and the coffee areas grew further and further inland.

From 1860 to 1885, upon the completion of several railways such as the Santos to Sao Paulo line, the interior forests of southeastern Brazil came under assault, with widespread cultivation of cash crops such as coffee. The plantations spread in all directions from the Campinas. Facilitated first by slave labor and later by immigrants from Italy, Spain, and Portugal the coffee *fazenda* prospered. By the end of the nineteenth century, more than 7.5 million acres of forests had been removed to grow coffee. Coffee growing and railroad construction bore devastating effects on the forests and the landscape. With the building of railroads, the interior of Brazil was opened up for human settlement and for the growth of plantations and cattle ranching. By the end of the nineteenth century over six thousand kilometers of railroads had been built in the forested region of southeastern Brazil. For the rest of the country, this came later, in the twentieth century. Railroads stimulated further landscape transformation, as they provided access to areas which in the past had been impenetrable. With this new access to the interior, deforested acreage on the coastal areas of Brazil was no longer viewed lucratively for further capital investment, and thus no attempts were made to undertake intensive agriculture. Instead, new virgin lands were sought. This meant that primary forests came under further pressure of removal. Railroads engendered deforestation not only by their construction; the running and maintenance of railroads required large volumes of wood. Crossties had to be replaced every six years, and for every kilometer of track fifteen hundred crossties were utilized.

Asia

By the nineteenth century, colonial rule was firmly established in most parts of Asia with the exception of China and Japan. Asia was carved up by Great Britain, France,

Netherlands, Portugal, and Spain. The trade in commodities, precious metals, cash crops, textiles, porcelain, etc. continued as it had for centuries. In certain areas, such as India, Southeast Asia, China, and Japan, either under colonial control or under indigenous rule, deforestation of the landscape persisted to meet colonial demands for wood or for cash-crop cultivation. Arable land increased substantially, and by the end of 1880, 187 million acres were under cultivation in south Asia and Southeast Asia. All these lands came from primary forest removal (Williams 1990).

Deforestation in India under British colonial rule was not a new phenomenon. Indian landowners prior to the arrival of the British were already transforming the landscape for agricultural production, such as the growing of cotton, pepper, sugarcane, indigo, etc. Exotic woods such as sandalwood and ebony were sought under the local land-use practices. The arrival of British colonial rule continued this land practice but geared it to meet the needs of an export trade (Williams 1990; Richards and McAlpin 1983; Tucker and Richards 1985). Imposition of taxes and revenue assessment by the British colonial government had the effect of encouraging land-clearing to pay for the levies (Williams 1990; Mann 1998). The lucrative cash-crop export trade further encouraged the destruction of primary forests to meet the global demands. Cash crops were sold at world-market prices amounting to five times more than the prices of food crops. As a consequence, for example, more than six million acres of forests southeast of Delhi were removed specifically for these purposes. Deforestation of the area also caused other outcomes that we have witnessed in the past. Salinization, water-table losses, and soil erosion were experienced. Substantial ecological changes followed as a result. In the Doab, loss of the forests raised land temperatures and increased the strength of the hot winds, the *lu* (Mann 1998). A great drought was experienced in 1837–1838 following years of dry weather. Considerable erosion also took place as a result of the loss of forest cover. Such climatic changes had consequences for soil fertility. This state of affairs led in the end to making this area of India agriculturally worthless as crop yields declined throughout the nineteenth century (Mann 1998).

The export of commodities and cash crops required transportation provided by the building of railways. Railway construction here as elsewhere utilized a significant amount of wood for railway ties. By the end of the nineteenth century, the British had constructed nearly 100,000 kilometers of railroads, from a total of only 7,678 kilometers in 1870. Forests were removed not only for railway construction. The teak forests of the western Ghats were a source of supply for Royal Navy shipping during the early part of the nineteenth century, and especially during Great Britain's rivalry with France.

By the nineteenth century, Southeast Asian trade was fully penetrated by the European powers, which had by this time established colonial control in various parts of the region. The British continued their empire from India to Burma, the Malayan archipelago, and Borneo. The Dutch focused on Indonesia, while the French were in Indochina. The Philippines remained with the Spanish until later in the century it came under the control of the United States.

The Dutch colonization of Java led to extensive landscape transformation, such as the degradation of the forests, exacerbated primarily to meet the production of exports such as sugar, coffee, indigo, and tobacco. This cash-crop production required the increased felling of trees to generate arable land for cultivation, for building construction, and for fuel to process sugar, tobacco, and coffee. Annual production of teak logs grew from an average of 16,700 logs during 1733–1765 to 145,000 logs during 1837–1865 (Boomgaard 1988, 66, 77).

Such economic practices and forest policies also occurred in other parts of Southeast Asia, where Great Britain was the colonial power. Malaya, Burma, and British Borneo witnessed the exploitation of their tropical forests by the British colonial administration to meet imperial economic ends and for shipbuilding. British companies searched for teakwood in Burma, and according to Rush (1991, 41) such quests "would play a part in Britain's annexation of Burma." Following the annexation of parts of Burma in 1826, British companies supported by the British colonial administration proceeded to exploit the wood resources (particularly teak) of Burma, and to clear lands surrounding the Irrawaddy River for rice cultivation (Adas 1983; Bryant 1996; Williams 1990). The need for wood was prompted by a shortage of hardwood for shipbuilding experienced by the British since the eighteenth century. The scale of deforestation to meet rice-growing needs was rapid. In 1850, approximately ten million acres of forests existed in the lower delta basin. By 1920, only several hundred hectares remained. Population levels grew from 1.5 million in midcentury to more than 4 million by century's end. Such scale of deforestation and the search for more land for rice cultivation even fostered the removal of coastal mangrove swamps. This practice continues to this day, where in most parts of the periphery, with dwindling energy and land availability, mangrove-forest removal becomes the last resort for the rural poor.

Deforestation in Burma had several consequences. In the lower delta it led to increases of malaria for the inhabitants. The loss of trees also meant that other forest products relied on by the local population were curtailed. This meant an increase in economic hardship, which prompted agrarian unrest. The destruction of the mangrove forests also meant the loss of local fishery stock, which exacerbated the poverty of the peasants, who relied on fishing as a source of income. Poorly designed embankments and irrigation systems also led to severe soil deterioration (Adas 1983). Besides soil deterioration, forest removal also led to the destruction of habitat of wildlife such as tigers and elephants, along with the loss of species of trees and plants.

In Malaya, the forests were cut to cultivate rubber plantations and for access to tin mines. To facilitate the establishment of a rubber-plantation economy, the British government encouraged the local rulers to sell their land and provided tax incentives to the new owners to plant rubber for export.[1] As in Java, the local Malay rulers, merchants, and traders were also intimately involved in the extraction of the forest and forest products and relied on the local aboriginal peoples for the extraction process (Kathirithamby-Wells 1998). Borneo was administered by a chartered company, the British North Borneo Company—much like the Dutch East India Company—which oversaw the management and harvesting of trees. The forest and forest products were in many cases the main source of revenue for the British North Borneo Company (Lian 1988; Kathirithamby-Wells 1998).

In Thailand, which was not colonized by the British, British timber companies also penetrated teak forests in the nineteenth century.[2] In the early twentieth century, Thai forests were granted as concessions to foreign companies from Great Britain, France, and Denmark. This manner of economic activity and penetration occurred also in the Philippines, except that the Philippines was first colonized by Spain in the sixteenth century and later by the United States in the late nineteenth century. By the end of the eighteenth century, the colonial government opened Philippines up to world trade, which led to forest removal for wood and for the cultivation of cash crops such as sugarcane (Roth 1983). Following the pattern in the Madeiras, the refining of sugar required a tremendous amount of wood as an energy source. The island of Luzon was severely deforested as a consequence of the fuel needs of the sugar refiners. Other islands of the Philippines also suffered deforestation. The islands of Mindanao and Samar specialized in the growing of abaca, which provided the raw material for cordage. Besides cash-crop cultivation, ranching, like elsewhere, was also undertaken, with much impact on the forests.

During colonial Spanish times, it is estimated that forests covered about 70 percent of the Philippines or about 19.5 million to 20.9 million hectares (Bautista 1990, 67). According to Tucker (1988, 223), following the control of the Philippines by the United States, the country was transformed after the turn of the twentieth century from a wood importer to a wood exporter. Prior to this, wood imports, primarily of Douglas fir and coastal redwoods, were shipped from the Pacific Northwest of America. The pace of cut increased following America's governance of the Philippines (Rush 1991; Tucker 1988). The Forest Law of 1904 formed the blueprint for the modernization of the logging industry in the Philippines. In it, U.S. colonial policy established the close relationship between the Bureau of Forestry and largely foreign (British and Spanish) and domestic timber companies. Besides this, a number of laws such as the Public Lands Act of 1902, the Mining Law of 1905, and Executive Order No. 27 of 1929 facilitated concentration of private ownership of land and opened up the forests for private commercial exploitation. According to Tucker (1988, 223), huge profits were made along with an increasing rate of depletion of the tropical forests of the Philippines. As with other parts of Southeast Asia, the local elites were active participants in this accumulation process. Such a heightened pace of activity led the transformation of Philippines from a wood importer to a wood exporter, and after the turn of the twentieth century, we find it exporting about 40 million board-feet per year of wood. By 1913, this volume had increased approximately threefold to about 112 million board-feet. The pace continued and the exploitation of the country's natural resources by foreign concerns (especially by the United States) transformed the Philippines into a colonial dependency of the United States

(Bello 1982). Annual rates of deforestation were at least 140,600 hectares per year from the 1920s to the 1930s. This rate increased by another 30,000 hectares per annum in the postindependence era (Bautista 1990, 69).

By the end of the nineteenth century, China had a population of nearly four hundred million people. Deforestation in China was mainly an indigenous affair since the European settlements only penetrated the coastal areas of China (Cox 1985). The high level of population—besides the needs of the industrial base and the extravagant consumption of the royalty and elites—led to severe transformation of the landscape. Arable land had to be secured for a rural-based economy. Forested slopes of south and central China were targets for the generation of arable land, mainly for rice farming (Williams 1997; Marks 1996, 1998). Commercial food production was not the only basis of deforestation in China. Wood was required for public buildings and homes, and as an energy source for heating and the manufacture of commodities such as porcelain, etc. By the early nineteenth century, however, wood as a source of fuel was no longer available in southern China and there was a problem with supply and frequent shortages (Adshead 1979; Marks 1996). State-sponsored land clearing, as in British India, was a feature in the deforestation process. Throughout the eighteenth and nineteenth centuries, China embarked on a total destruction of its forests (Murphey 1983; Williams 1997). This has led Williams (1997, 179) to suggest that by the beginning of the twentieth century, China was "an impoverished peasant society, subsisting in a treeless, denuded and eroded lunar landscape." Murphey (1983, 111), however, has even suggested that by the mid-nineteenth century most of China's forests were removed and what was left was destined for further internal consumption.

The consequences of these acts of forest removal to meet population growth and the reproduction of an economy can be clearly seen in southern China. Denuded land led to invasion by weeds, which in the long run made the land uncultivable, and over the long term the forest did not have the opportunity to regrow. Marks (1996) reports that constant burning by the local peasants further ensured the failure of the trees, and thus the forest, to return. Without cover, soil erosion often followed, which led to further loss of soil fertility, in addition to the flooding and siltation that often followed soil erosion. One other outcome of forest removal that Marks (1996, 1998) has noted is the loss of wildlife. In this case, the tiger was threatened.

DEFORESTATION IN THE TWENTIETH CENTURY

With the European control of the world's economy firmly in place by the twentieth century, the process of the accumulation of capital on a world scale continued unabated. Regardless of which country, civilization, or region has dominated the world economy, the accumulation process has persisted with the removal of the forests just as it had for the last five thousand years of human history. Despite political independence achieved in most regions, such as in Africa and Asia in the first half of the twentieth century, political liberation did not usher in a period of respite for the forests. The indigenous governments, spurred on by the call of nation-building, combined with neocolonial policies on the part of the West resulted in a modernization trajectory that has meant the emulation of the developmental strategies of the West (Chew and Denemark 1996). Such developmental policies required the utilization of the natural assets of each country, including the availability of cheap labor to meet five-year developmental plans designed locally or in conjunction with expatriate help. The developmental plans included dam building, resettlement schemes involving poor urban populations, highway and railway construction, monoculture cash-crop cultivation, minerals and oil extraction, and import substitution/export industrialization. The end result has meant the massive transformation of the landscape to meet developmental goals spurred on by the availability of foreign aid and private commercial loans to foster respective nation-state developmental policies and the business operations of local elites and multinational corporations. The latter have become the successors to the early trading companies that were discussed in the previous chapters.

The enduring core–periphery relations persist along the same lines as always; in the twentieth century the periphery, that is, the Third World, became the resource extractive base. Some manufacturing activities were also conducted in these locales to meet the global accumulation process. Resource specialization—dependent on climate, topography, natural assets, or simply available low-cost and willing workers—has been the determining factor in the global division of labor. The three regions of the Third World (Africa, Asia, and Latin America) now provide the resource base for the accumulation processes of Europe and the United States of America.

In the late twentieth century, the exploitation of the forests remains a worldwide global process. The assault occurs across the continents of Europe, Africa, North America, Asia, and Latin America. In Latin America, Africa, and Asia, the tropical rainforests have been systematically cut to meet accumulation and consumption needs that are both local and global in nature. In the Pacific Northwest of North America, the ancient forests remain under assault. Of the twenty million acres that existed in western Oregon and Washington in the nineteenth century, only 12 percent or 2.4 million acres were still standing in the early 1990s. About eight hundred thousand acres are protected in parks and wilderness areas, with the remaining open for harvesting. The rate of cut was around seventy thousand acres in the 1980s, though this has been reduced somewhat on public lands with the Northwest Forest Plan. What remains, however, is very fragmented. The level of fragmentation in the early 1990s is much worse than in the Amazon rainforests, according to a NASA scientist (Egan 1992).

Either operating independently or in collaboration with local timber and lumber concerns, multinational companies based in Japan, Europe, Canada, the United States, and even Malaysia have been active in harvesting the global forests (Petesch 1990; Colchester 1994). Deforestation of the world's landscape reached epic proportions toward the end of the twentieth century; the world's forests have shrunk by nearly half, from 6 billion hectares to 3.6 billion hectares, since eight thousand years ago (World Commission on Forests and Sustainable Development 1999). The rate of harvest of wood is growing about 1.5 percent annually, amounting to 75 million cubic meters with about 15 million hectares of forests cleared per year, largely in the tropics (World Commission on Forests and Sustainable Development 1999; Noble and Dirzo 1997).

With the growing world population, now at 6 billion and expected to reach 9.5 billion by 2050, it is estimated that an additional 10 million hectares of land will be needed, which will be cleared from the forests for agricultural purposes to feed the growing number of people. According to the World Commission on Forests and Sustainable Development (1999), forests have virtually disappeared in twenty-five countries, eighteen others have lost more than 95 percent of their forests, and eleven countries have lost 90 percent. Such losses on the world scale lead to the decline of biodiversity. It is estimated that 12.5 percent of the world's 270,000 species of plants and about 75 percent of the world's mammals are threatened by the decline in the world's forests.

REGIONAL IMPACTS

Africa and Latin America

In Latin America and Central America, from 1950 onward, export cash crops (sugar, coffee, dyes, bananas, etc.) were cultivated, continuing the trend that started with the first European contact in the sixteenth century. Cattle ranching was also undertaken to supply meat to Europe as a consequence of the latter's loss of arable land due to urbanization and industrialization, notwithstanding population increases. Clearing these lands was facilitated by refinements in technology such as bulldozers, tractors, and chainsaws. Forest removal was the norm; virgin forests were the prime targets. Such relentless push from the 1950s forward led to more than half of the forests of Central America being removed by the mid-1980s. The impacted areas covered the countries of Nicaragua, Guatemala, Panama, Honduras, Costa Rica, Belize, Mexico, and El Salvador. The most devastating act, which continues even today, is the deforestation of the tropical forests of the Amazon. Under sharpest attack are the lowland tropical forests that covered 61 percent of the Amazon river basin (Denevan 1973; Fearnside 1982; Goodland and Irwin 1975; Moran 1982; Marchak 1995).

Highway construction in the Amazonia from the 1950s to the 1970s, starting from the Belem-Brasilia Highway and the Transamazonia highway, provided the opportunity for resource extraction, cattle ranching, and population migration. More than two million people moved as a consequence of the Belem-Brasilia Highway (Williams 1990). Direct resettlement schemes, such as the Independent Colonization Project and the Rondonia Colonization Scheme, facilitated the assault. Shifting cultivation following forest removal added further stress to the landscape, along with cattle ranching and mining to meet the needs of an export market in Europe and North America. During this period the tropical rainforests in Venezuela and Peru came under assault as well.

Deforestation rates differ across regions. In Latin America, debt servicing, multinational activities, ranching, and international organizations have been responsible for the destruction of the Brazilian rainforests and other parts of Central and Latin America (Barbosa 1993; Hecht and Cockburn 1990). The rates continue to increase in Latin America but seem to have stabilized for Africa and Asia. Latin America and the Caribbean, which together account for 28 percent of the world's forest today, sustain an extremely high deforestation rate. Between 1990 and 1995, Central America lost about 2.2 percent of its forests per year. The Caribbean came quite close to this amount, averaging about 2 percent per year. Over a fifteen year period, the whole region lost about 10 percent of its total forest area.

Africa, which holds 15 percent of the world's total forest, had by 1995 lost half of it. These tropical, moist forests are found mainly in coastal West Africa and equatorial central Africa. Six countries (Cameroon, Central African Republic, Congo, Equatorial Guinea, Gabon, and Zaire) house 60 percent of the total. Earlier in the twentieth century, French colonial demands in Sahelian Africa engendered cash-crop cultivation (peanuts and cotton), which extended deforestation through a policy of taxation much like what the British implemented in India (Thomson 1988). Instead of intensive farming techniques, extensive farming was practiced. Such economic practices did not bode well for the land. In the latter half of the twentieth century, deforestation tended to be on a grander scale in West Africa, where not only land cultivation was at play but also the export of tropical woods. The rate of deforestation has led to Nigeria, a major timber exporter in the 1960s, now becoming a net importer of wood products.

In the late twentieth century, the Amazon rainforest continues to come under assault. The rate of deforestation was about five million acres per year throughout the 1980s. In the early 1990s, there were some reductions to an average of four million acres per year. However, this shortfall seems to be only temporary, as the rate has increased again since 1995, when a recorded eight million acres were removed. Besides the Amazon rainforest of Brazil, in the Southern Cone, the temperate forests of Chile are also increasingly being removed. In place, tree plantations for pulp and paper manufacturing have taken root. The tendency to replace virgin forests with tree plantations seems to be the operating principle by both the public and private sectors in the late twentieth century as a response to the dwindling available forests, and as a way of dealing with the sustainability issue raised by deforestation.

Southeast Asia

The tropical rainforests of Southeast Asia continued to fall under the axe in the postindependence period (Hurst 1990). It is the region that has witnessed one of the most severe deforestation rates in the late twentieth century, thus reflecting the long, continuous process of Culture's degradative impact on Nature. Whereas in the past colonial administrations approved the timber concessions to meet the needs of the imperial power, we find in the postindependence period the replacement of this authority by the region's newly independent states to meet nation-building needs. The replacement of authority and rationale did not lead to a change in the trend of exploitation of the tropical rainforests. With nation building as the task of the state, coupled with Western-driven models of development as an operating principle—enhanced and encouraged by the presence of Western expatriates, bilateral and multilateral aid agencies from the North, and local bureaucrats and intellectuals with training in the imperial countries—the aim was to build upon the strength of an export-oriented economy (including natural-resource extraction and cash crops) established during the colonial era, and at the same time to industrialize. Continuing this nation-building process, the construction of an infrastructure for a modern society was also undertaken with much vigor. As a consequence of the global division of labor and the peripheralization of these countries, they had very few choices other than to take the Western developmental path. This type of Western development strategy bore its costs on the environment (in our case on the tropical rainforests), especially when land had to be cleared for commercial agriculture and dams had to be built to provide hydro-electric power for urban and industrial uses. Notwithstanding these developmental policies, population increases have also been rapid in the postcolonial era. The population increases were complemented by an increase in urbanization from the mid-twentieth century onwards that further exacerbated an already stressed ecosystem. From 1630 to 1930, Southeast Asia remained the least urban region of the world. However, from 1950s onwards, urbanization has been progressing at an alarming rate. In the 1940s, there was no city with a population over one million. At the end of the twentieth century, there are three cities (Manila, Bangkok, and Jakarta) that have populations exceeding ten million.

The players in this postindependence development drama did not change much. The local political elites (royalty in some countries) and merchants along with the foreign companies continued to participate in forest extraction. The former colonial administration was now replaced by bilateral and multilateral agencies providing policy advice and loans. With independence, more economic space was provided to the local merchants, financiers, and political elites to expand their economic activities and further their accumulation of capital. Southeast Asian states increasingly expanded their powers by playing a very major economic role in the development of the economies of these countries. Regardless of whether a country was governed by elected politicians or by military leaders, from country to country one could see the exploitation of the forests to further the economic interests of the ruling elites. Mechanized timber harvesting was utilized and encouraged, followed by the loosening of timber regulations (Williams 1990; Rush 1991).

Beginning with Indonesia, which already had Java deforested during the Dutch colonial period, we find the largest timber concessions being located in Sumatra and Kalimantan. In East Kalimantan we find joint ventures such as Weyerhauser (United States) and P.T. Tri Usaha Bakti (a company owned by the Indonesian armed forces), which in the 1970s extracted wood from a concession of some six hundred thousand hectares. Other transnational companies, such as MacMillan Bloedel (Canada), Georgia Pacific (United States), Scott Paper Company, Mitsubishi (Japan), Sumitomo (Japan), Shin Asahigawa (Japan), and Ataki (Japan), also penetrated and maintained timber operations in Indonesia (Hurst 1990, 34). Political patronage as an operating principle during President Suharto's years led to the concentration of the Indonesian timber and lumber industry; the five hundred or so timber concessions (approximately 50 percent of Indonesia's forest covers) were operated by fifty conglomerates that not only do the extraction of the timber but are also involved in downstream wood manufacturing such as plywood production, sawn lumber, and furniture manufacturing. Plywood production has increased. From 29 plywood mills in 1980, 130 mills blossomed by the early 1990s. The pulp and paper industry has raised capacity by more than six times between 1988 and 1997, investing over eight billion dollars (Leahy 1999). The close working relationship between capital and the state is exacerbated further by the lax payment of reforestation taxes, and by illegal logging and forest poaching (Chew 1995b; Dauvergne 1997). Recent reports indicate that the Indonesian wood industry is dependent on illegal logging for more than 70 percent of its consumption of domestic raw timber. For example, in 1998, Indonesia consumed more than seventy-eight million cubic meters of domestically sourced logs, while the official supply figure was only about twenty-one million cubic meters. In other words, a large portion of these logs come from the clear-cutting of natural forests at unsustainable rates. The cost in terms of forest loss was an average of 1.5 million hectares a year between 1985 and 1997, leaving only 20 million hectares of quality production forest left, according to estimates (Leahy 1999).

In neighboring Malaysia, the pattern is similar. Logging, once important for the economy in West Malaysia, is now on a decline following intensive cutting from the 1960s to the early 1980s. The focus of the timber industry for the 1990s shifted to the states of Sarawak and Sabah, which now account for 70 percent of Malaysia's output. East Malaysia is the world's largest exporter of tropical hardwood logs (Belcher and Gennino 1993, 25). In Sarawak's case over half of its forests (4.9 million hectares) have been awarded as forest concessions without any requirement for competitive bidding. Unlike other Southeast Asian countries, Malaysia has complex laws surrounding foreign ownership of timber concessions. Therefore, foreign input into timber and lumber extraction on the whole is via the point of sale or through the lines of credit extended by foreign trading companies (mainly Japanese) to local logging concerns, or through holding companies and joint ventures.

It is clear that the Malaysian federal and local governments, either in their pursuit of developmental policies or in enhancing the interests of some elected politicians, have driven the trend and intensity of the exploitation of the tropical rainforest of Malaysia. Advised and encouraged in the postindependence era by multilateral agencies such as the World Bank and the Asian Development Bank to target timber as a major export for developing the country, the state in Malaysia has, in concert with local economic elites (and in some cases with foreign corporations), proceeded to exploit this natural resource without much restraint, ignoring even the plight of its aboriginal peoples (Hurst 1990; Gedicks 1993).[3]

The trend of exploitation of the tropical rainforests of Southeast Asia differs little in the Philippines. In certain aspects the situation is more acute, with 80 percent of the ancient forests already logged, especially during the administration of Ferdinand Marcos (1964–1985). In a pattern similar to that of East Malaysia, the state supported foreign and local timber and lumber interests to pursue the development of the wood industry. Since independence, a major foreign presence in the Philippines are timber and lumber concerns from the United States and Japan, such as Georgia Pacific Corporation, Findlay Millar Company, Boise-Cascade Corporation, International Paper Company, Weyerhauser Corporation, Nisso Iwai, Marubeni, Yautaka Mokuzai, Mitsui Bussan, Mitsubishi Shoji, C. Ito, Sumitomo Ringyo, Meiwa Sangyo, and Yuasa Sangyo. Along with local financiers these concerns have considerable influence on the Philippines forest industry. Return on investments during the 1970s was between 25 percent and 30 percent a year, and Hurst (1990) suggests that through various subsidies and tax benefits a company would not need to make further reinvestment.

The procedure for awarding timber-license agreements is similar to the East Malaysian practice. During the Marcos Administration, awards were based on connections with the president or someone in the Wood Industries Development Board (Hurst 1990, 187). Stumpage fees and taxes were minimal, amounting to one peso per hectare (0.5 cents U.S.) and 1.5 peso for every cubic meter harvested. An export tax was also collected, but nonpayment was quite rampant as the level of illegal export of logs (amounting to 60 percent of the legal exports) between the 1960s and 1970s was quite high (Bautista 1990, 69; Hurst 1990, 190). Timber-license agreements were awarded to private concerns, and from the postwar period to the early 1980s, these agreements amounted to more than 70 percent of the classified forest lands of the Philippines. The length of the timber-license agreement was usually twenty-five years, with the opportunity for renewal. Most often, renewal is not requested as the area has been deforested to such an extent that it is not financially feasible to continue logging operations.

Thailand, which deftly played one colonial power against another, was never colonized. This does not mean, however, that foreign penetration of its forest industry did not take place. With forest products being Thailand's main export during the early part of the twentieth century, foreign companies participated in the widespread logging activities. The state, of course, supported this penetration, sold rights to log the forests, and collected duties on the export of logs. The trade in logs focused primarily on teak, and foreign operations were quite dominant until 1960. With the vast deforestation over the course of the twentieth century and the ban on all logging of ancient forests in 1989, foreign concerns such as Shell Corporation have shifted to tree plantations to produce wood chips.

With no colonial history, the economic space in the timber and lumber industry is wider for local companies. However, the rate of exploitation of Thailand's forests is just as rapid, if not more so, and very few virgin forested areas remain. Such starkness can be seen, as the country now has only 22 percent of its landmass classified as forest (Belcher and Gennino 1993, 39). What is clear is that the national origin of companies does not impact on logging practices or on environmental protection. As in the Philippines and East Malaysia, illegal logging in Thailand is rampant and led the state to declare a total ban on logging in 1989. High rates of deforestation have caused Thailand to import wood to meet its domestic needs since the late 1980s. The last five decades of postindependence assault on the tropical rainforests of Southeast Asia have produced a scarcity in hardwood for countries like Thailand and the Philippines. The specter of shifting from wood exporting to wood importing has emerged for Thailand and the Philippines in the 1990s.

Deforestation Outcomes

If we examine areas that are still classified as ancient virgin forests, it is very clear that the Southeast Asian region as a whole has undergone massive deforestation. According to the United Nations Economic and Social Commission for Asia and the Pacific (1990), dramatic deforestation in Asia occurred between 1950 and 1976, when approximately four million hectares were cleared per annum with a high percentage in Southeast Asia. For Asia, from the 1970s onwards rapid deforestation continued to meet global and regional needs. Deforestation rates averaged 1.1 percent per year between 1990 and 1995. In India, the forests of the Himalayas are being removed (Guha 1989). Indonesia topped the list with about 600,000 hectares per year, followed by Thailand (252,000 hectares), and Malaysia (250,000 hectares), with Burma (105,000 hectares) and the Philippines (91,000 hectares) at lower levels. These levels have now been exceeded; in 1992, Indonesia was clearing 900,000 to 1,300,000 hectares a year, while Burma was cutting between 600,000 and 1,000,000 hectares a year (Rush 1990, 35; Belcher and Gennino 1993, 34). It is also evident that private timber and lumber enterprises, whether of foreign or local origins, have played significant roles in the deforestation of the tropical rainforests of Southeast Asia (Petesch 1990, 3). Other contributing causes include poverty leading to further degradation of the environment for fuel; poor agricultural techniques (for example, swidden farming) of the aboriginal peoples; the opening up of forested areas for settlement or for other economic activities such as mining, cattle ranching, and cash- or food-crop production; the population increase, leading to pressure for more arable land; and urbanization.

Some of these causes have been widely documented and circulated by development agencies such as the World Bank, the Asian Development Bank, the Food and Agricultural Organization of the United Nations, and even the Brundtland Commission (1987). No doubt on the surface one can find evidence for these causes leading to deforestation and environmental degradation. However, we would suggest that some of these causes are outcomes rather than contributing factors. Indeed, poverty, urbanization, cash-crop production, and cattle ranching in Southeast Asia have caused deforestation. But these "causes" are socioeconomic conditions generated by other social, economic, and political trends that have been engendered by state policies, the global division of labor (such as the production of cash crops for core consumption, as in cassava production in Thailand for animal feed in the EEC, palm oil and cocoa production in Malaysia for core consumption, etc.), development policy advice and funding from multilateral and bilateral aid agencies, and foreign or local capitalist enterprises seeking to further the capital accumulation process. Other contributing factors that have engendered deforestation are state-sponsored rural-development migration programs and infrastructure development, such as the building of dams and roads. These kinds of modernization activities, which occurred under the slogan of rapid rural development and modernization, were extremely popular from the 1960s to the 1980s among states in Southeast Asia and among the multilateral and bilateral development agencies. The migration programs of Indonesia and Malaysia led to the clearing of forests for human settlement. Primarily these programs were to ease population pressure, to provide land for the landless, and to foster development. Such migration programs are not new—previously the Dutch and British colonial administrations moved the population during the early part of the twentieth century, of course for differently stated reasons than the above. For the Dutch it was to diffuse resistance to land acquisition by the colonial power, and for the British it was to combat communist insurgents and reduce the number of recruits to the communist insurgency in postwar Malaya. The migration programs that have now been implemented under the two national governments are capital and resource intensive and utilize bulldozers to clear the forests for farming. Most of the Indonesian migrants (a projected three million) are being moved to Kalimantan and Irian Jaya, parts of Indonesia that have abundant land with remaining tropical rainforests. To fund this program the state has borrowed U.S. $637 million from the World Bank (Hurst 1990, 27).[4] The bank is also funding the rehabilitation of deforested areas next to the cleared migration sites. In Malaysia's case, the opening up of rural areas and the awarding of land ownership to the landless rural poor is undertaken by the Federal Land Development Agency (FELDA). FELDA's land schemes are primarily to provide land for the production of cash crops such as cocoa, rubber, and palm oil.[5] Between 1966 and 1990 a total of 1,301,300 hectares of forest was cleared for FELDA land schemes. In addition, during the same period other non-FELDA land-development schemes cleared 2,473,470 hectares (Government of Malaysia 1966, 1981, 1986). It is clear that such land development programs have exacerbated the deforestation process in Southeast Asia, and consideration has to be given to how effective these programs have been for the "new settlers." Hurst (1990, 66) has indicated that the policy of FELDA in Malaysia to encourage the growing of cash crops such as palm oil, rubber, and cocoa might be misguided, as it links small-scale farmers to the vagaries of international pricing for these products. Thus, when prices fall (as in the mid-1980s) the Malaysian farmers are hurt severely. Notwithstanding all this, it is clear that the Western modernization strategy pursued in Southeast Asia and in the rest of the world has been the overall cause for this regional deforestation and global deforestation.

From this brief sweep of Southeast Asian developmental and natural-resource extraction strategies, we can safely say that the intensity of clearing the forest to meet socioeconomic ends has been massive during the postindependence decades.[6] In country after country we witness the rapid pace of cutting and clearing the forests. The danger the world faces is that 23 percent of the world's rainforests are located in Indonesia, Malaysia, Papua New Guinea, Philippines, and Thailand. It is also widely cited that the rainforests of Indonesia and Malaysia have among the greatest diversities of species found on planet Earth. The rainforest is being cleared in Indonesia at a rate of about 900,000 to 1,300,000 hectares a year, without much regard for the loss of this biological diversity. It was estimated in 1992 that there were only about 94,000,000 hectares of forests left in Indonesia. Additionally, Indonesian mangrove forests, estimated at about 4,200,000 hectares in 1982, were reduced to around 1,000,000 hectares by 1985, of which 500,000 hectares are now protected (Hurst 1990, 3). In neighboring Malaysia, the rate of deforestation remains high. The total annual cut, measured in 1989, was about 800,000 hectares per year (*Utusan Konsumer* 1989). In 1981, it was estimated that West Malaysia had 2,900 square kilometers of virgin rainforest remaining, and that the logging industry cut about 149,000 hectares annually (half of which is classified as virgin). In East Malaysia, in the 1960s, there were about 4,500,000 hectares of forest in Sabah, accounting for 60 percent of the land area, whereas in Sarawak the forest covered about 9,400,000 hectares (76.5 percent) of the land area. The rate of deforestation in Sarawak is quite extensive: 2,820,000 hectares were cut between 1962 and 1985. Thus, in twenty-three years, 30 percent of Sarawak's forests were cut. Peninsular Malaysia and Sabah are expected to lose their forests by the late 1990s, with Sarawak following suit in the twenty-first century (Belcher and Gennino 1993, 25).

In the Philippines, more than 80 percent of its virgin forests have been lost (Rush 1990, 42). According to Bautista (1990, 67–68), the country had almost 30,000,000 hectares of forest area during the Spanish colonial times; by 1988, 6,500,000 hectares were left. From 1934 to 1988 old-growth forests shrank from 13,600,000 hectares to about 990,000 hectares. From the time of American colonization to the postindependence period, most of Philippines' dipterocarp forests were cleared. Byron and Quintos (1988, 431) indicate that between 1995 and 2000 it is expected that Philippines' virgin forests will be completely logged. The situation is similar in Thailand, where the state claimed in 1985 that 115,000 square kilometers of forests remained, or about 22 percent of the land area. Over the preceding twenty-four-year period, the country had lost 31 percent of its forests, which in 1961 covered 53 percent of the land area or 273,628 square kilometers (Hurst 1990, 239). Such intense cutting has resulted in the country's importing wood to cover its domestic needs.

Besides deforestation, other side effects of this process should be addressed; of great concern are soil loss, erosion, intermittent flooding of lowland areas during the monsoon season as a consequence of losing forest cover, and the deliberate use of forest fires for land clearing. Malaysia, Indonesia, Thailand, and the Philippines have experienced these conditions. For example, the island of Java has been losing 770 million tons of topsoil annually, which reduced the crop yield by 5 percent per annum (Hurst 1990, 4). In the Philippines, the Ministry of Natural Resources has estimated that one-third of the land area of the country is severely eroded, and twenty-five of its provinces suffer severe erosion of over 50 percent of their land area. Soil loss is followed with sedimentation problems that affect river flows and harbor depths. This has often led to flooding, as it has in the West Malaysian states of Terenganu and Kelantan, and to the need to dredge harbors, as in Sarawak in East Malaysia. Indonesia, Philippines, and Thailand have also experienced these conditions. Concomitant with this are the effects on fish stocks either in the mangrove areas of Malaysia, Indonesia, and the Philippines or in the Gulf of Thailand. The sedimentation issue has also affected the acidity levels of the rivers, and in turn this has affected aquaculture and fishery stocks in the coastal zones.

Forest fires have been regular occurrences. They have happened as a result of lightning or other natural circumstances. In the late twentieth century, however, some forest fires are set in Indonesia and other countries deliberately to clear virgin tropical forests for cultivable land (Vishvanathan 1997). In 1997, a series of forest fires raged over more than 10 million hectares of tropical forests in Indonesia, causing devastation and losses amounting to nearly $9 billion (*International Herald Tribune* 1999). Smoke from the fires reached atmospheric levels high enough that it even affected surrounding countries such as Malaysia and Singapore. Losses suffered by these countries are estimated at about $4 billion in the areas of tourism, transportation, and public health. During certain periods of this ecological crisis, schools and airports in the impacted areas were closed due to the excessive smoke. Native wildlife was impacted by loss of food sources and habitat. A global response to provide fire-fighting equipment was made. Naturally, such widespread devastation also impacted on the Indonesian economy, which was already under stress. Towards the end of this millennium, it seems that the practice of deliberately setting forest fires to clear lands for plantations continues even today (*International Herald Tribune* 1999). With the economic crisis experienced in Southeast Asia in the late 1990s, this trend will continue as Indonesia attempts to overcome its economic problems through the export of cash crops and timber.

Besides erosion, another urgent and important threat to the environment as a consequence of deforestation is the loss of biodiversity. As we have stated, the tropical rainforests of Southeast Asia have among the greatest diversities of plant and animal species on this planet. The loss to date has been drastic. In West Malaysia, which has 7,900 species of flowering plants, 207 species of mammals, 495 species of birds, and 250 species of freshwater fish, we find that logging has put on the verge of extinction 61 species of mammals and 16 species of birds; a further 130 species of mammals and 148 species of birds are on the vulnerability threshold. Regarding the number of primates lost as a result of deforestation, Hurst (1990, 121), for example, has indicated that some species, including the siamang, the white-handed gibbon, and the dusky leaf-monkey, suffered population losses of more than 50 percent between 1958 and 1975. No doubt these losses have increased by now. The threat to wildlife diversity also occurs in Indonesia, Thailand, and the Philippines. Some species, including the Philippine monkey-eating eagle, have been reduced to only six hundred pairs. The elephant populations in Thailand, as well as the orangutan and the Java and Sumatran rhinos in Indonesia, are being threatened (Belcher and Gennino 1993, 17).

Along with the outcomes catalogued above, we find that the aboriginal peoples in these countries are systematically being dislocated or relocated from their traditional abode (Gedicks 1993). In country after country, from Thailand to the Philippines, the aboriginal peoples or ethnic groups have been uprooted so that the assault on the rainforests can continue or dams can be built. In each case the social conditions are different but the outcomes are the same: the transformation of these peoples from leading independent, self-sustaining lifestyles to leading lives dependent on government assistance or low-wage work in the timber industry. Malnourishment is one of the outcomes of this dislocation. In Sarawak, 72 percent of the rural Dayak children were malnourished in 1979, compared to the Asian average of 3.2 percent (Hurst 1990, 87). The assaults on the forests have led to radical protests, supported by environmentalists, by some of the aboriginal groups in the region (such as the Consumers Association of Penang, Sahabat Alam Malaysia). The Penan tribal protests in Sarawak, East Malaysia, became one of the most publicized events in the region. The state response to this has been to utilize the law to imprison protesters.

Russia

Throughout the 1980s and 1990s, with mounting pressure from environmental groups in North America, the search for wood has moved to Siberia, and Canada's boreal forests are also now seen as potential resource bases (Goto 1993, 6; Chew 1995b; Massa 1999). Russia has 23 percent of the world's forest areas and contains 11 percent of the world's biomass. The Russian forests have more than 55 percent of the world's growing stock of coniferous species (Nilsson and Shivdenko 1997). Decreases of these species were quite significant from 1983 to 1993. Overharvesting by clear-cutting has impacted on the forests located in the European part of Russia. In the Russian Far East, only large trees were harvested, and the removal rate was between 45 and 65 percent of the forested area. More than eight million hectares were cut between 1965 and 1988.

With the fall of the former Soviet Union and the transition to an open economy geared for profits, Russian forest resources have been made more available for the world wood markets. As a consequence, multinational corporations from the core have penetrated the Russian timber industry. For example, Japanese multi-nationals such as Mitsubishi and Sumitomo in conjunction with Svetlaya have a joint-venture deal with Hyundai Corporation of Korea to exploit the Siberian taiga (Goto 1993, 6). A total of sixty-six joint ventures were signed during the 1990s (Massa 1999). According to Nilsson and Shivdenko (1997), Russian forests were severely impoverished during the period 1961–1993. Impoverish-ment also means endangered species. About six hundred plant species (trees and bushes) and about fifty animal species are threatened or endangered. As else-where, aboriginal groups are also impacted. There are approximately thirty dif-ferent groups, totaling about two hundred thousand persons, living on hunting, fishing, and reindeer breeding in the affected Russian areas. Because of increased ecological degradation, their traditional activities are being threatened.

WOOD MARKETS AND JAPAN

In the late twentieth century, the United States and Japan have replaced England and France as the two main centers of wood consumption, and consequently of wood accumulation. At this stage, the dynamics of wood exploitation occur in parts of the world system that are the most amenable to the accumulation of cap-ital. Consequently, we find extensive wood operations by American and Japanese multinationals in North America, Latin America, Asia, Siberia, and West Africa. In Japan, whose wood resources were severely exploited in the past, global ef-forts are made now by the *sogo shoshas*[7] to maintain a constant supply, relying on foreign sources (Nectoux and Kuroda 1990, 27). In Southeast Asia, Japanese *sogo shoshas* have managed to penetrate the tropical rainforests (Chew 1993; Dauvergne 1997). Japanese dependency on imported wood has risen from 5.5 percent in 1955 to 55 percent in 1970 and 66.5 percent in 1986 (Nectoux and Kuroda 1990, 27). With a more bountiful indigenous wood supply, the United States has been more fortunate than Japan. However, this has not stopped multi-national operations of U.S.-based corporations from exploiting the temperate and tropical forests in the periphery of the world system.

Core–periphery relations have also engendered differential accumulation rates and fostered the ascendance of core states in the world system, as evidenced by the history of accumulation via timber and lumber operations. In the nineteenth century, England and France seemed to benefit. In the late twentieth century, besides the United States, we find Japan exhibiting a global accumulation strategy in the timber and lumber sector of the world economy. The activities of Japan's *sogo shoshas*, including wood import levels, reflect the intensiveness and breadth of the capital accumulation process.

For nearly two decades, Japan has been the world's major tropical-hardwood importer. Total volume of tropical-wood imports into Japan amounted to 29 percent of the world trade in tropical hardwoods in 1986 (Nectoux and Kuroda 1990, 5). These wood imports have been to meet Japan's timber and paper requirements. The high levels of consumption reflect an age of exuberance much like what the United States experienced in the postwar era (Devall 1993).[8] Housing starts, measured as number of houses per thousand persons, in Japan have overtaken those in the United States since 1986, considering that the United States has a higher population level than Japan. Between 1970 and 1990, on average, Japan built 1.458 million homes per annum; the United States built 1.609 million homes per annum (Yorimitsu 1992, 3). When one averages these figures in terms of houses per one thousand persons, Japan was building almost twice the number of houses per thousand persons as the United States (12.5 houses per thousand persons in Japan against 7.1 houses per thousand persons in the United States) (Yorimitsu 1992, 3). This clearly shows a level of rapid economic development and capital accumulation. In the area of paper products, Japan over the last thirty years has emerged as the world's largest importer of forest products, second-largest producer of paper and paperboard, third-largest producer of pulp, and second-largest consumer of paper in the world after the United States (Penna 1992, 1). Total consumption of paper and paperboard has increased more than 548 percent during the last thirty years (Penna 1992, 3). Because of this, Penna (1992, 1) has suggested that Japan's "paper and timber markets dominate the world forest product trade. Its paper industry is the world's largest importer of woodchips and the third largest importer of pulp from the world's markets."

Because of its proximity to Southeast Asia, Japan has treated the region as its own wood yard, just as England did with British North America in the eighteenth and nineteenth centuries. During the 1980s to the early 1990s, it imported unfinished logs from Southeast Asia averaging eleven million cubic meters per year, or about 38 percent of its total volume of log imports (Kato 1992, 95). In fact, the highs occurred in the early 1980s, with volumes above 12 million cubic meters and as high as 19 million cubic meters (50 percent of total log imports into Japan) in 1980. Malaysia, Indonesia, and Papua New Guinea were the main contributors (Mori No Koe 1993, 8). By the end of the 1980s this slowed not because of a drop in consumption, but because of a diminishing resource and the bans enforced by the states in Malaysia, Indonesia, and the Philippines on the export of raw logs.[9] Other contributing countries to the wood industry of Japan were Canada, the United States, the former Soviet Union, and New Zealand.

Other finished wood products also showed a high importation volume, depending on the type of product. Volume of lumber imports from Southeast Asia between the years 1980 and 1990 varies somewhat from a low of 10 percent of total lumber imports into Japan in 1980 to 18 percent in 1989 (Kato 1992, 96).[10] This cannot be said for plywood-import volumes. Between 1980 and 1990, Indonesia, Malaysia, and Canada topped the list. In fact between 1989 and 1990, Indonesia provided 97 to 98 percent of the total volume of plywood imported into Japan, with Canada and Malaysia making up most of the remaining.

What does this mean for the Japanese corporations involved in this sector of the world economy? Investments in the pulp and paper industry have grown steadily from 1.8 billion yen in 1980 to 4.7 billion yen in 1989 (Graham 1993, 39). This also includes direct investments in other parts of the world economy, such as in the United States, Canada, Chile, Indonesia, Thailand, Malaysia, Portugal, New Zealand, Australia, and Papua New Guinea (Penna 1992, 29–30). According to Penna (1992, 18) the concentration of the share of production in the industry has increased dramatically over the last thirty years. In 1960, the top ten Japanese pulp and paper corporations produced 61 percent of the pulp, 64 percent of the paper, and 42 percent of the paperboard in Japan. By 1990, the top ten corporations produced 83 percent of the pulp, 74 percent of the paper, and 57 percent of the paperboard. These corporations are the Japanese conglomerates (*keiretsu*—of which the most widely known are Mitsubishi, Mitsui, Sumitomo, Sanwa, and Fuyo) and the general trading companies (*sogo shosha*—of which the major ones are Marubeni, C. Itoh and Company, Kanematsu, Nissho Iwai, Nichimen, and Tomen), which play a key role in the organizing of crucial foreign wood-chip and pulp projects and in the trading of these wood commodities (Penna 1992, 21).

Production volumes came under regulation by the Ministry of International Trade and Industry (MITI) in the 1980s to force a restructuring and to reduce overcapacity. Following the lifting of the regulatory production levels, the pulp and paper industry is still required to submit plans if production is to be increased. Such regulatory practices have led to the concentration of companies in terms of volume of production. Such a high volume of consumption naturally leads to an immense power on the part of Japan's pulp and paper industry over its peripheral producers in the world, such as in Southeast Asia. Tadem (1990, 23) has indicated that the prices of Asian wood products are determined via the buyer's market, of which Japan has the major role. Japan has been labeled an economic imperialist by some of the Southeast Asian wood producers, for it levies higher import taxes on finished wood products than on logs. For example, a 20 percent import tax is levied on plywood imports against 0 percent for unprocessed logs. It has been suggested that this is to protect Japan's labor-intensive wood-manufacturing industry (Tadem 1990, 24). Other features of economic reactions by Japan to coordination by Southeast Asian producers (Council of Southeast Asian Lumber Producers Association) to control supply have been met with the slashing of the prices of logs and lumber, "to break up any wood and forest product cartel in the region" (Tadem 1990, 25). It is safe to say that such actions are hardly unique in the history of the world system. One needs only to look at past British and American policies toward their colonies or at those countries under their scope of economic penetration/domination.

At the end of this long historical sweep of the relationship between Culture and Nature and its consequences, one often asks the question of whether the history of the human community's relations with Nature ever reflect some sensitivity to Nature that is not exploitative and degradative. Are there moments in the human community's consciousness, philosophies, and even practices that move beyond an exploitative relation? It is to these questions that we turn in the final chapter.

NOTES

1. By the 1950s, tin and rubber were the major exports of Malaya, and rubber plantations covered approximately fifteen thousand square kilometers of Malaya.

2. The penetration of Thailand was made during the reign of King Mongkut, who signed the Bowring Treaty with British companies. The teak was paid for with opium (Rush 1991, 16).

3. The state in Malaysia is by no means unique in its actions towards its aboriginal peoples. During the nineteenth century, the U.S. state, in a series of strategies to develop its western frontier, also ignored the plight of its Native Americans. One can also find instances of this in Canada's historical development.

4. The program provides two to four hectares of freshly cleared forests, a basic wooden house, a supply of foodstuffs for one year, basic tools for farming, seed supplies for the first harvest, fertilizers and pesticides for the first three years, free transportation to the sites, and the provision of infrastructure such as roads, schools, and clinics (Hurst 1990, 22). From the development point of view and not from an environmental point of view, such a program would enhance the development of the country, provided the country has an autocentric economy. However, if the country has a dependent economy, the financial resources allocated for such a program would benefit the core in terms of products imported and interest paid on the development loan.

5. Almost similar in awards, FELDA schemes provide to each family 4.5 hectares of cleared land, a basic house that has a 6.5 percent annual mortgage interest rate, along with the supporting infrastructure such as roads, schools, and clinics.

6. The world's remaining rainforest is being lost at about one hundred acres a minute; about 1.8 percent of total rainforest area is lost per year (Porritt 1990).

7. The *sogo shosha* is a trading house that provides financing of timber operations in the peripheral areas, the credit for processing, and sales. There are approximately eighty-five hundred of these trading houses in Japan; however, the fifteen largest ones handle the majority of timber imports into Japan (Nectoux and Kuroda 1990; Penna 1992).

8. It should be noted that due to mounting pressure from environmental groups in North America, a more consistent supply of wood resources is now being sought in Siberia (Goto

1993, 6). Since the 1960s, Japan has imported wood from the former Soviet Union, though because of an antiquated production process marred by outdated machinery, Soviet wood could not compete against wood from North America in terms of quality. With constriction of wood supply, Siberia is now seen as a potential resource base. Mitsubishi and Sumitomo in conjunction with Svetlaya are in a joint-venture deal with Hyundai Corporation of Korea to exploit the Siberian taiga (Goto 1993, 6).

9. Indonesia started to restrict the export of logs in 1979, while Malaysia took more than a decade later to ban the temporary export of logs in January 1993 and to lift the ban with quota-level exports in May 1993 (Mori No Koe 1993, 6).

10. Canada heads the list in terms of overall increasing volume of imports between 1984 and 1990.

Chapter Nine

Ecological Consciousness and Social Movements among *les ancients et les modernes* 2700 B.C.–A.D. 2000

Throughout world history and across geographic space and time, as we have witnessed in the previous chapters, the excessive degradation of Nature by different cultures, civilizations, and kingdoms has been unceasing as the processes of urbanization, population growth, and capital accumulation spiral upwards. The point of focus for analyzing such a trajectory rests on the relationship of Culture with Nature. Deliberations have therefore focused on how to overcome this exploitative relationship of Culture on Nature. Such discussions have generated discourse on how to modify cultural consumptive practices via initiatives such as reducing consumption, recycling, promoting environmentally sustainable resource use, etc., as well as on the more intellectual need of reorienting our philosophical and social attitudes to Nature. With regard to the latter, for some, the basis for such a reorientation can be sought in the progressive Western Enlightenment philosophies and their associated rationalization processes. Within these frameworks, it is claimed, are the seedbeds for such a reorientation, and the process of modernity as it spirals onward can usher in the stage of ecological consciousness that will supplant the other, prior stages that modernity has disintegrated or made obsolete (Eder 1996; Bergesen 1995). Such positions underscore the theme that the Nature–Culture relationship and the development of ecological consciousness are conditions that human social formations arrived at progressively over historical time, and that across world history the rationalization processes have become increasingly attenuated to a progressive context whereby human realization becomes more cohesive, less alienated, and more symbiotic with Nature. This modernist position, because of its social evolutionary developmental stance, assumes that for prior eras and for other cultures, civilizations, and kingdoms, such tendencies are not exhibited because the progressive march of Enlightenment history and philosophies had not occurred, and therefore the conditions for such a reorientation had not developed.

Unfortunately, this myopic modernist position fails to account for what has transpired in world history. If we survey the narratives on Culture and Nature and the ecological context in which human systems are reproduced, we most often find that these have recurred at least throughout the last five thousand years in the philosophical and cultural deliberations of urbanized civilizations, kingdoms, and states. These considerations therefore are not solely the domain of Western Enlightenment philosophies, nor are they exclusively the franchise of civilizations and empires that have been considered the seedbeds for the emergence of Western civilization, such as classical Greece and Rome. By no means can we attribute the dialogue on Nature–Culture relations singularly to the processes of the modern era or to the stage reached in societal evolutionary developmental sequences or to only Western Enlightenment "progressive" philosophies. If it is the case that the dialogue on Nature–Culture relations has occurred in various cultural circumstances across geographic zones and across various time periods, the intent of this chapter is to discuss the existence of such ecological consciousness and awareness through time and space, and to suggest that Culture–Nature relations have always preoccupied human communities throughout world history, and especially during periods of severe ecological degradation.

NATURE–CULTURE RELATIONS ACROSS WORLD HISTORY

Southern Mesopotamia: 2700 B.C.

As early as 2700 B.C., the Nature–Culture relationship was one that appeared to permeate the cosmologies of communities. In the early Mesopotamian kingdom of Uruk, circa 2700 B.C., we find a characterization of this Nature–Culture discourse in the epic of Gilgamesh (Perlin 1989; Harrison 1992; Glasser 1995; Torrance 1998). In this narrative, as depicted in twelve tablets that were recovered from the library of the Assyrian king Ashurbanipal, the relationship of Culture with Nature is portrayed as one whereby Nature is subjugated by Culture to meet its needs. Pari passu, Culture is also considered a part of Nature. For the latter, the conception is that Nature is intrinsic to the human identity, and that Nature forms and conditions the worldview of the human being. This Culture–Nature link for personal and social identities becomes weaker and weaker as Culture progressively evolves and develops along socioeconomic and technological dimensions. Such a duality of relationship appears not only in the worldview of the southern Mesopotamians, but this underlying theme of the Nature–Culture relation recurs through time and across cultural and civilizational groupings.

These Nature–Culture themes are characterized in the recovered tablets. In tablet one, Gilgamesh, who is supposed to have ruled Uruk as its fifth king for 126 years, was seen by the sky god, Anu, as having abused his kingly powers. In turn, Anu called on the goddess, Aruru, to create a second image of Gilgamesh in the form of Enkidu. Enkidu is the personification of the human being's intrinsic connection with Nature, for he lives among Nature and feeds with the animals. In this short section of the tablets, it is clear that the writer, Sin-leqi-unninni, concentrated on the evolution of Culture away from Nature and the development of the human being's control over others (humans and Nature), as exemplified by the oppressiveness of Gilgamesh to his people (Torrance 1998). To restore this tilting away from Nature, Enkidu, who has all the qualities of Nature—the naturalness of the human—is created in an attempt to replace what has been lost or forgotten, and to underscore the relationship between Culture and Nature as complexes that mutually condition each other. We suspect that what is implied is the notion that through Nature, the human person realizes his/her true nature and that any systemic unbalancing or breaking from this relationship generates uncertainty and human excessiveness:

When Anu the sky god heard their lamentation
he called to Aruru the Mother, "Great lady: You, Aruru, who created humanity,
create now a second image of Gilgamesh: may the image be equal to the time of his heart.
Let them square off one against the other, that Uruk may have peace."

When Aruru heard this, she formed an image of Anu in her heart.
. . . .
In the wilderness she made Enkidu the fighter; . . .
His whole body was covered thickly with hair, his head covered with hair like a woman's;
the locks of his hair grew abundantly, like those of the grain god Nisaba.
He knew neither people nor homeland; . . .
He fed with the gazelles on grass;
with the wild animals he drank at waterholes;
with hurrying animals his heart grew light in the waters.

(Torrance 1998: 97)

The harmonious relationship between Nature and Culture can never be maintained in view of different cultural complexes' systemic need for materialistic reproduction and their social evolutionary capacities of increasing consumption. In the paragraph that follows, Nature's power and immensity are acknowledged, and Culture's materialistic utilization of Nature is depicted in the form of a trapper who intrudes on Enkidu's landscape. Materialistic consumption also provokes cultural conservation impulses. At this point in the epic, we have the introduction of ecological conservation as a theme underlying the Culture–Nature relationship. Noting the time period that the epic is supposed to cover, it would be one of the earliest pieces of literary evidence of ecological conservationism in world history (Torrance 1998, 96).

The social appropriation of Nature through the cultural exploitation and transformation of Nature is reflected in the trapper's action to change the naturalness of Enkidu with the help of other social agents, in this case, Gilgamesh and a courtesan. Enkidu, seduced by the courtesan, thus loses his naturalness and is enticed to leave the forest for civilization and urban living. For in this latter setting, the materialism of life is reproduced. To reproduce the materialistic needs of urban communities, as we have discussed in earlier chapters, Nature's bounties have to be extracted and utilized. Urbanization thus exemplifies the "culturalization" and "denaturalization" of the human community.

My son, in Uruk lives a man, Gilgamesh:
no one has greater strength than his.
In all the land he is the most powerful; power belongs to him.
Like a shooting star of Anu, he has awesome strength.
Go, set your face toward Uruk.
Let him, the knowing one, hear of it.
He will say, "Go, Stalker, and take with you a love-priestess, a temple courtesan,
let her conquer him with power equal to his own.
When he waters the animals at the watering place,
have her take off her clothes, let her show him her strong beauty.
When he sees her, he will come near her.
His animals, who grew up in the wilderness, will turn from him."

Then, five stanzas later:

After Enkidu was glutted on her richness
he set his face toward his animals.
Seeing him, Enkidu, the gazelles scattered, wheeling:
the beasts of the wilderness fled from his body.
Enkidu tried to rise up, but his body pulled back.
His knees froze. His animals had turned from him.
Enkidu grew weak, he could not gallop as before.
Yet he had knowledge, wider mind.

Turned around, Enkidu knelt at the knees of the prostitute.
He looked up at her face,
and as the woman spoke, his ears heard.

The woman said to him, to Enkidu:
"You have become wise, like a god, Enkidu.
Why did you range the wilderness with animals?
Come, let me lead you to the heart of Uruk of the Sheepfold,
to the stainless house, holy place of Anu and Ishtar,
where Gilgamesh lives, completely powerful,
and like a wild bull stands supreme, mounted above his people."
                                                    (Torrance 1998, 96–98)

The "culturalization" and "denaturalization" processes continue, and the former is triumphant in the end. This is reflected in Enkidu's losing his fight with Gilgamesh and becoming his ally. As the epic continues, the kingdom of Uruk is transformed during the rule of Gilgamesh, and he proceeds with an ambitious plan to build the city. The Culture–Nature link continues to widen and weaken as urbanization proceeds and population increases follow. We discussed these trends and tendencies in chapter 2. To fuel the urbanization process requires the exploitation of Nature. At this early stage, the wide-scale use of wood to meet material reproductive needs was the norm. Such materialistic requirements, which have not subsided even today, resulted then in the deforestation of the cedar forests near the kingdom of Uruk (Perlin 1989; Harrison 1992; Glasser 1995). This deforestation process, undertaken by Gilgamesh with Enkidu, underscores explicitly the degradative relationship of Culture upon Nature, with Gilgamesh never recognizing the natural limits. Despite warnings from the guardian of the cedar forests, a demigod named Huwawa/Humbaba, the destruction of the forests continues unabated in spite of wails from the cedars after Humbaba is decimated (Perlin 1989, 38). This exploitative relationship with Nature, and in this context with the forest, is repeated over and over in world history and is related to us in the various expressions of ecological consciousness across time and geographic space.

South Asia: 2500 B.C. and Beyond

Accounts and interpretations of the Culture–Nature relation of the early Indus civilization are quite limited. Primarily, this is because the language of the Harappans has not been deciphered, and what has been asserted has been based on archaeological finds and the interpretation of terracotta seals (Lal 1997; Possehl 1996). It seems that the Culture–Nature relation underlying the Harappan civilization, as we delineated through a materialistic examination of their socioeconomic practices in chapter 2, is one that is underscored by a high degree of exploitation and degradation by the former on the latter. On one hand, our historical examination shows this is the case. On the other hand, it has been suggested that the Harappans related harmoniously with their environment, based on the terracotta seals that have been uncovered indicating the Harappans' veneration for other living beings (Chapple 1993a). Without any Harappan literary sources depicting intellectual and religious ideas or practices, deciphering the ecological consciousness of the Harappans is a limiting endeavor that relies mostly on what the seals reveal.

Accounts of ecological ethics derived from the post-Harappan age do minimally indicate the Harappans' ethical consideration of nonviolence to other living beings. Tahtinen (1991), in a review of Indian values and ecology, has suggested that the moral concept of *ahimsa* has inputs from two sources: *sramanic* and Vedic. In other words, this central moral concept is, according to Tahtinen (1991, 214), an outcome of the reciprocal influence of two different ideologies. The *sramanic* interpretation is devoted to ensuring that human activities that are mental, vocal, or physical in nature do not cause any suffering to other living beings. However, the Vedic version is not as comprehensive in terms of nonsuffering, for it has a provision for animal sacrifices. In this light, the *sramanic* version cannot be considered to have been based on the Vedas, and is supposed to be pre-Vedic in origin. If this is the case, it is assumed that the *sramanic* interpretation had its roots in the early Indus civilization (Tahtinen 1991, 212).

If the above representation of *ahimsa* is convincing, it would imply that the ecological consciousness of the Harappans was guided by a duality of relationship with Nature and other living beings. One dimension covering a materialistic and degradative aspect relates to Nature in an exploitative fashion. Such a relationship contributed to civilizational collapse, as I contended in chapter 2. The other dimension, as explicated in the moral concept of *ahimsa*, if Tahtinen's (1991) interpretation is valid, underlines an ecological ethic of a harmonious connection with Nature. The available accounts of Harappan civilization, unfortunately, do not allow us to project further than this in terms of the ecological consciousness of this early civilization. In a contemporaneous sense, these ecologically ethical tendencies parallel that of southern Mesopotamia, as I have discussed previously.

In a later period, between 1500 B.C. and 900 B.C., these early ecologically ethical tendencies are explicated in the collection of Vedas: Rig-Veda, Sama-Veda, Yajur-Veda, and Atharva-Veda. The narratives found in these Aryan writings seem to have affinities to the polytheism of the early Greeks (Callicott 1994). The outlook on Nature—in our case, ecological awareness—is one of perceiving it as part of a continuous spectrum of relationships constitutive of the human being (Bilimoria 1998). Social and moral order thus are perceived as correlates of the natural order. In the Atharva-Veda, this normative value is awarded a deeper ecological bent through the interpretation that everything in the universe has its own worth and therefore is morally significant.

Thus, the Vedas underline the conception that everything is part of a unity, and that every part is constitutive of a whole. In terms of Nature, we note such reflections in the Atharva-Veda:

Untrammeled in the midst of men, the Earth,
adorned with heights and gentle slopes and plains,
bears plants and herbs of various healing powers.
May she spread wide for us, afford us joy!

On whom are oceans, river, and all waters,
on whom have sprung up food and ploughman's crops,
on whom moves all that branches and stirs abroad—
Earth, may she grant to us the long first draught!

To Earth belong the four directions of space.
On her grows food; on her the ploughman toils.
She carries likewise all that breathes and stirs.
Earth, may she grant us cattle and food in plenty!

Then, eleven stanzas later:

All creatures, born from you, move round upon you.
You carry all that has two legs, three, or four.
To you, O Earth, belong the five human races,
those mortals upon whom the rising sun
sheds the immortal splendor of his rays.

We pick up again eleven stanzas later:

Earth is composed of rock, of stone, of dust;
Earth is compactly held, consolidated.
I venerate this mighty Earth, the golden-breasted![1]

(Torrance 1998, 125–26)

Ecologically speaking, this underscores the interconnectedness of everything within a complex, interdependent system. What this leads to is awareness that every element of Nature has a purpose to fulfill and unfold within the larger schema of the grand order of things.

What is recognized from this is that Nature cannot be controlled nor disrupted, as the order of things has its proper functioning. The state of equilibrium has to be maintained, and competition over resources could disrupt and generate an imbalance. What is thus prescribed is an attitude of mutual respect and reciprocity. A sense of caring for the environment was thus the consequence. In the Vedas, the writings reveal a significant understanding of biodiversity, the relationship between living beings in Nature, the natural order of equilibrium and dynamism, and the costs of disrupting the ecological dynamics and flows. The universe is viewed as a meaningful moral order and human beings thus have a moral responsibility to it. From this, therefore, flows the idea of noninjury or nonviolence to all living beings.

This ecological awareness as explicated in the Vedas later became subdued in the Upanishads (Jacobsen 1994; Smith 1989, 1990; Zimmerman 1987). With the cultural evolution of Indian society into a hierarchical order matched with materialistic reproduction of life that is couched within public sacrifices and prayers, the ecological consciousness so prevalent in the Vedas was overtaken by these pragmatics (Bilimoria 1998, 5). As we have witnessed in southern Mesopotamia, the culturalization process subdues the Culture–Nature relations.

The ecological awareness of the Vedas emerged again three hundred years later in the sixth century B.C.—a period of economic downturn—in Jainism (Callicott 1994; Torrance 1998; Chapple 1993a, 1993b). Contemporaneous with the rise of Buddhism, Jainism's basic precept is that every entity in the world possesses *jiva* or a sentient principle. This sentient principle is distinguished by consciousness coupled with energy. In this context, consciousness occurs in every living thing throughout the universe, and thus every living thing has a degree of sentience. This degree is contingent on the state of embryonic evolution, with its most developed form found in the adult human being. Such an orientation suggests that the living world is formed of interconnections, and therefore the human being is related to Nature inasmuch as Nature is related to the person (DeBary 1958, 60–61).

With this acknowledgment of the principle of sentience in every living being, Jainism thus operates with five practical ethics: *ahimsa*, which is noninjury; *satya*, meaning truthfulness; *asteya*, fostering not to steal; *brahmacharya*, or sexual restraint; and *apigraha*, meaning nonpossession. Of the five ethics, two (*ahimsa* and *apigraha*) have implications for human relations with Nature. This is especially so when *apigraha* means a rejection of material possessions, which has implications for the ecological balance. For world history has shown that the continuous culturalization of life requires a continuous materialistic consumption of natural resources. This is especially so when social life is organized within a hierarchically social division of labor. It would seem, therefore, that Jainism engenders an awareness that is supportive of an ecological worldview (Chapple 1993a; Callicott 1994).

Contemporaneous to Jainism is Buddhism. Unlike Jainism, however, Buddhism does not subscribe to the notion of *jiva* for every living being (Chapple 1993b). In spite of this, both share the basic precept of not taking life. As such, the Buddhist code of ethics parallels Jainism's ethics such as self-control, abstinence, patience, contentment, purity, and truthfulness. These ethical positions thus warrant the human being to treat animals and plants accordingly. The individual is directed to have reciprocal compassion towards other life forms and to take the "middle path" away from polar extremes. Such positions developed an ecological awareness that led to the conservation and protection of other lifeforms.

As Buddhism spread from India to Asia, including to Japan and China, from A.D. 1 onwards, we witness the transformation of these Buddhist ethical principles to laws and practices protecting the treatment of animals (Callicott 1994; Bilimoria 1998, Chapple 1993a; Torrance 1998). Even in the India of its birth, we find the establishment of laws requiring the kind treatment of animals by the Emperor Asoka (274–232 B.C.) when he converted to Buddhism. These laws included restricting the consumption of meat products, the curtailment of hunting, and the establishment of roadside watering facilities for animals. Such practices spread in areas where Buddhism took root, especially in Asia. During the sixth century A.D., in China, the Buddhist monk Chi-i convinced over a thousand fishermen to give up their occupation and at the same time purchased three hundred square miles of land as a conservation area for animals (Chapple 1993a). Later on, in A.D. 759, this practice of conservation was followed by Emperors Suh-tsung and Chen-Tsung during their reigns. In their cases, it was the establishment of conservation areas for fish. In Japan, around A.D. 675, Emperor Temmu Tenno ordered restrictions on meat consumption and curtailed the use of certain hunting devices. This was also the case during the reign of Emperor Shomu Tenno in A.D. 741, when hunting and fishing were prohibited. These practices were continued by his daughter, Empress Koken, in several decrees.

China: Sixth Century B.C. and Beyond

Notwithstanding the spread of Buddhism into China after Christ, Chinese philosophy prior to this was already deeply imbued with a sense of ecological awareness (Chan 1963). Foremost in this regard is the ancient philosophy of Lao Tzu (Chan 1963) or the *Tao te ching*, which appeared around the sixth or perhaps the third century B.C. (Chan 1963, 137). For in the *Tao te ching* are the explications of an ecological consciousness that parallels the ecological worldviews of late-twentieth-century deep ecologists (Callicott 1994). The Taoists were "antiurban, antihumanistic, and antibureaucratic. . . . They were bioregionalists" (Callicott 1994, 67). Central to the ecological worldview of the Taoists was the Tao, meaning the "way." The "way" represents our harmonious relationship with the world, where Nature and human are one, and the process is one of cyclic change and reciprocal processes. In terms of human relations with Nature, it is to be exemplified by the ideal of action, *wu-wei*. Within the Taoist context, *wu-wei* means not-doing, and it directs the practitioner to flow with the Tao rather than against it. The person is to follow Nature, and in so doing the individual is fulfilled (Chan 1963). From the ecological point of view, human action thus has to conform with the flow of events and things rather than against the current. In this respect, the flows of Nature are not interrupted, nor are excessive energies channeled towards disrupting the flow. Such a worldview thus conforms to the natural order of events and in this fashion engenders human actions that are not catastrophic to Nature, which in this case incorporates human beings as well.

Extending this ecological awareness even further, Chuang Tzu, in his treatise on Nature circa fourth century B.C., believed in nourishing Nature and enjoying it (Chan 1963; Torrance 1998). The human individual is part of Nature and derives his own nature from it. Thus, Nature is a constant flow, and the individual follows the streams of Nature. From this comes the source of the individual's essential learning and ultimate truth. In certain fashion, such a position embraces a level of ecocentricity that had not yet been encountered in the ancient philosophies. It would be safe to state that the philosophy of Chuang Tzu reveals the naturalization of the human being, and that the culturalization process is dependent on and finds its basis in Nature.

If Taoism celebrates Nature and its central place within the cosmology of the human individual, Confucianism emphasizes the social beingness of the human individual. In a certain sense, it is anthropocentric in inclination towards the proper action by the individual. Notwithstanding this anthropocentricity, two early Confucian scholars, Mencius (371–289 B.C.) and Hsun Tzu (298–238 B.C.), were quite ecologically aware in terms of understanding the relations of the human individual with Nature. The latter, in his treatise on Nature, noted the symbiotic relations of human communities with the natural environment (Chan 1963). To Hsun Tzu, Nature has to be treated with respect and its bounty has to be used wisely without excessiveness (Chan 1963, 116–17). For to do otherwise would mean bringing disaster to the human community. Mencius's writings, too, share this line of reasoning. For in his work, there is consideration of land management as an important facet of the reproduction of the human community (Hughes 1997).

In practical terms that are beyond the realm of philosophies and worldviews, the idea of ecological balance was already in place by the period of the Sung Dynasty, from A.D. 960 to 1275. A number of environmental decrees were passed during the northern Sung period (Edmonds 1994).

Greece and Rome: Sixth Century B.C. and Beyond

During a period when the *Tao te ching* was considered in China, Greek philosophers such as Pherecydes, Pythagoras, Philolaus, and Empedocles were also developing philosophical discourses on human relations with Nature (Torrance 1998; Hughes 1975, 1994; Goldin 1997; Adams 1997; Westra 1997; Glacken 1967; Sessions 1985). In this context, Nature and humans are seen as one, and in the case of Pherecydes, the earth was contemplated as a living organism. Especially for the Pythagoreans, all living things share common elements, and the process of life is a constant movement of recycling. If this is the case, as with the Vedas there is the belief in the protection of life, and that means killing is prohibited. To the Pythagoreans, human consumption does not require the killing of living beings, and the food necessary for consumption can be obtained without having to end a life. Milk, cheese, fruits, wine, oil, and honey form the basis of the diet suggested, and our following it safeguards Nature and the souls of the living beings. Plato was, of course, influenced by the Pythagoreans in his deliberations on human relations with material reality, though he emphasized human society more than Nature throughout his work. Philosophical and religious viewpoints of this type were also shared by the Greek Stoics, who deemed that all living beings participate in the cycles of life, and that Nature is a harmonious process filled with sentient living things.

This ecological assertion of the harmonious system of Nature became hierarchical with the deliberations of Aristotle. Though he believed in the hierarchical order, he would not agree to the senseless taking of the lives of other living things, and he did suggest that each living thing exists to reproduce the needs of others. What this means is that there are interconnections. To Hughes (1994, 63), Aristotle should be given credit "for introducing ecologic considerations into scientific literature."

Despite this awareness of the ecological interconnections and the place of the humans in the web of life, the early Greeks, like the Mesopotamians and the

Harappans before them, were also degraders and exploiters of their ecological niches. Deforestation and land degradation were very prominent during the height of classical Greek civilization, as explicated in chapter 4 of this book. It seems, therefore, that these two dimensions of human relations (conservation and degradation) with Nature pervade world history, and most often it is the latter dimension that predominates. We hear this echo resoundingly during the era of the Roman period, though guided by a more pragmatic approach.

Influenced by the Greeks, Roman philosophers and poets viewed Nature as hierarchically ordered (Hughes 1975, 1994). As such, Nature was interconnected, and the human had an active role in being its caretaker. To this extent, Cicero and Pliny the Elder complained about the abuse of Mother Earth by the Roman way of life, and in a pragmatic fashion they advised the wise use of resources. But Roman attitude to the environment was often guided by a trend towards utilitarianism and practicality. What this means is that the land was seen as offering much to meet the needs of the Roman world. Numerous accounts were written about the land and its bounties.

Religious ideas further determined the relation with the land. Heavily influenced by animism, the Romans believed that spiritual powers were manifested in Nature. To this end, certain locations such as forest groves, mountains, and streams had special considerations for Roman life. Temples and shrines were built to signify such manifestations, and, on occasion, animal sacrifices were made. Clearly, in this context, the Romans veered away from the practices of the earlier civilizations that observed veneration for other living beings. Roman practices, as discussed in chapter 5, underlay the domineering relations with Nature that led to widespread ecological stress. Despite this tendency, the Romans did extend rights, *jus animalium*, to animals but not to the rest of Nature (Nash 1982). The pragmatic practicality orienting Roman relations with Nature continues to this day in our interactions with Nature and in our attempts to reproduce a materialistic lifestyle across cultures and geographic space.

Ecological Consciousness in the Modern World

Across geographic space and time, deliberations about the relationship between Culture and Nature continue. In the modern context, what has been deliberated by past civilizations and cultures—from the southern Mesopotamians to the Chinese—gets repeated and reconstructed vis-à-vis the changing ecological conditions. With the decline of Rome and the advent of Christianity, Nature did not fare well. According to Nash (1982, 17), the book of Genesis prescribed for humans their dominion over Nature and the right to exploit it without any restraint. This *lebenswelt* meshes with the Romans' view of Nature, as we have discussed in chapter 5. Such an anthropocentric stance was buttressed further by the ethics of René Descartes during the sixteenth century A.D., who deliberated that animals were insensible, and thus the nonhuman world was a thing/objectification

(Descartes 1901, 1983; O'Briant 1974). This furthered the distantiation between Culture and Nature and provided a framework that denaturalized Culture.

In contradistinction to the belief of Descartes and echoing the calls of Jainism and Buddhism more than twenty-three hundred years earlier, John Locke concluded otherwise. Locke reasoned that animals can suffer and to harm them would be morally wrong. Beyond just placing an ethical position on the harming or taking of animal life, Spinoza placed ultimate ethical value on the system as a whole (Sessions 1985). For him, the system is composed of interrelationships, with no hierarchical order in Nature. Thus, every living thing is connected to everything else, and everything has a right to exist and possesses its own worth. Recall that Spinoza's stance is reflective of the Indian Vedas.

The Spinozan theme of interrelationships was taken up by Henry David Thoreau (1972, 1982) in mid-nineteenth century North America. For Thoreau, there is an organic holism, whereby every living thing, for example, in the Maine woods, is a community of living beings. Hierarchical order does not exist for Thoreau, and every creature is better alive than dead (Thoreau 1972). He would further declare that human domination of Nature would be inappropriate. This emerging ecocentric bent is nudged along, without being explicitly so, by John Muir (1911, 1916, 1986). For Muir, everything in Nature has a value. The human does not hold court over Nature but is just a part of the living ecosystem. His essays reveal his veneration for each living thing in Nature, and the beauty and spirituality of the wilds of America. Hence, he promoted the defense of these wilderness places, such as the Hetch Hetchy Valley in Yosemite National Park.

The community of interdependent parts and the veneration of Nature that permeated environmental consciousness in India, China, and Greece from 1500 B.C., and in the works of Thoreau and Muir in A.D. 1860s continues to our current era. Foremost is the work of Aldo Leopold (1949), who insists that the human individual belongs to a biotic community of interdependent parts. From an initial utilitarian position towards Nature—the greatest good for the greatest number of people over the longest possible run—Leopold's belief shifted at the end of his life towards seeing the human individual not as a conqueror of the land but as a plain member and citizen. This position forces us to respect the other members of Nature, and to consider that humans therefore are only an element within a living ecosystem. In the final chapter of *A Sand County Almanac*, Leopold calls for a land ethic that treats the land with love and respect following our realization that we, along with other living things, belong to a community of interdependent relationships: "In short, a land ethic changes the role of *Homo sapiens* from conqueror of the land-community to plain member and citizen of it. It implies respect for his fellow-members, and also respect for the community as such" (Leopold 1949, 240). This land ethic, i.e., Culture's relations with Nature, is circumscribed within the ambit of respect for the stability and beauty of the land. It leads Leopold (1949, 262) to suggest that a land-use decision is right "when it tends to preserve the integrity, stability, and beauty of the biotic community. It is wrong when it tends otherwise."

The foregoing provides a glimpse of treatises on Culture's relations with Nature in the modern world. What this brief examination has disclosed is that ecological awareness during the modern era continues to focus on issues that have pervaded intellectual, philosophical, and religious inquiries over the ages. In short, the ecological discourse of the modern era is as old as the hills. In the later period of the twentieth century, deliberations on the Culture–Nature relationships have revealed attempts to integrate different cultural interpretations of the Culture–Nature relations, albeit continuing with the theme of a living ecosystem of which the human is only one component. Gary Snyder's works (1974, 1980, 1983) follow such lines when he attempts to develop a Culture–Nature relationship incorporating Buddhist and Native American principles. His interpretation of the Nature–Culture relation is one based on ecocentrism, wherein plants and animals are also "persons" and thus have a place and a right to exist and blossom. What this means is that other living beings are just as important in our environment, and that Nature has to be preserved for its own sake. Another example would be the work of Arne Naess (1973, 1984, 1985, 1986, 1987, 1989). Naess (1986) notes the parallel nature between his ecological philosophy (ecosophy) and those of Buddhism and Taoism concerning the human place in the ecosystem. A philosophy of understanding and underscoring ecological harmony or equilibrium forms his ecosophical framework. Naess reasons that if environmental supremacy and ecological harmony are to be maintained, then every living being has the right to live (Naess 1973). This suggests that each living being has a value in itself, leading to the notion of intrinsic (inherent) value. Realization of this, via the process of identification, leads to the development of the ecological self (1985). This defines a relationship with Nature that parallels what we have encountered in the Vedas and in Taoism, for example. At the hands of Naess, Nature receives ontological status.

Social Movements and Conservation Practices

The existence of ecological consciousness over the ages would also mean the appearance of environmental movements and conservation practices. By no means are environmental movements just a feature of the late twentieth century A.D., nor should they be viewed as *new* social movements of protest as some have posited (Castells 1998; Touraine 1985a, 1985b). Evidence of their emergence throughout the early parts of world history is available; of course, they were not as widely prevalent as they were in the late twentieth century.

Information on groups that embrace alternate life philosophies and practices prior to the Christian period is limited. We have some indication that the practice of vegetarianism by Pythagorean groups in classical Greece was a rejection of the dominant culture with its bloody sacrificial rites (Eder 1990). During the Roman expansionary period, there was recorded resistance by countermovements to Roman imperial ecological dispositions, especially by peasant farmers in the outlying areas of the empire (Grove 1990). Of course, even then there was the emergence of intellectuals whose works covered ecological conservation issues (as discussed in the previous sections), especially when there were increasing signs of environmental crisis in terms of wood supplies.

In the Christian era, and especially with European colonialism, we find after the seventeenth century an increasing awareness of the relationship between colonial rule and environmental destruction, leading to various legislative acts to protect the environments of the colonies. Starting in 1677 with the Mauritius, there was the stark realization that the chopping of ebony forests would have an ecological impact, and coupled with social unrest there was emerging resistance by intellectuals of the day (Grove 1990, 1992). French scientists such as Philibert Commerson, Pierre Poivre, and Jean Jacques Henri Bernardin de St. Pierre undertook initiatives to protect the environment. For them, responsible stewardship was viewed "as an aesthetic and moral priority as well as a matter of economic necessity" (Grove 1992, 43). Laws were passed for the Mauritius covering forest protection and water pollution from the indigo and sugar mills. French conservationism was followed by its British counterpart in Tobago and India. In addition, there are indications since the seventeenth century of social unrest and resistance to imperial ecological edicts and interventions from the farmers and peasants of Africa and Asia (Grove 1990).

Two centuries later, in America, we have the emergence of the environmental movement with the founding of the Sierra Club in 1892 by John Muir and his associates. Despite differences between the conservation and preservation wings of the U.S. environmental movement, it has continued, although with splits in philosophy and tactics. The emergence of radical environmentalism in the 1970s—with its critique of the mainstream movement as being "less responsive to grassroots demands for more rapid change in public policy, too bureaucratized and centralized, and too 'shallow' in ideology"—has changed the landscape of resistance to capital penetration and accumulation (Devall 1992, 51–52). Direct action in the form of ecotage, guerilla theater, demonstrations, and "bearing witness"[2] by radical environmentalists and ecowarriors are some outcomes of this resistance (Manes 1990; List 1993; Devall 1991, 1992; Foreman 1992). Ecological activism thus took on the theme "resist much, obey little" (Devall 1991, 138).[3] Radical environmental groups that have formed include, for example, Earth First!, Greenpeace, Sea Shepherd Society, Rainforest Action Network, and Rainforest Information Center. The latter two groups bring people together (via networks) who are interested in a common area, which in this case is the tropical rainforest. Their actions have focused on the importance of protecting native biodiversity and the human rights of the native aboriginal peoples living in the rainforests through demonstrations at annual meetings of the World Bank, stockholders' meetings of multinational corporations involved in rainforest extraction, and educational campaigns directed at segments of populations in North America, Europe, Japan, and Australasia, including appeals to governments.

In the late twentieth century and the early twenty-first century A.D., there are a plethora of environmental groups in all the regions of the world economy (Gottlieb 1993; Brulle 1996; Dalton 1994). They range from local groups acting to save their immediate environmental surroundings to those addressing national and global concerns. They range from groups protecting the forests to movements for environmental justice (Szasz 1994). Politically, there have been gains in terms of winning political power, especially in western Europe, where the Green Party has won seats in the German Parliament (Kelly 1994). Viewed from the long-term world-historical perspective, environmental movements always seem to emerge during eras of ecological degradation and crisis. As such, they rise and ebb like the tides, fluctuating as ecological recovery and degradation occur in phases as the materialistic reproduction of life spirals through world history.

ECOLOGY IN COMMAND AND THE PROBLEM OF LEARNING

As we come to the end of this inquiry, a question emerges from this brief sweep of world ecological practices: why is it that across cultures and through these long centuries of ecological awareness, the predominant point of orientation for human practices has always focused on the maximal utilization and degradation of the environment, instead of pursuing the view of the human community as belonging to a biotic ecosystem of interdependent relationships guided by a "land ethic"? In this book, several world historical processes (accumulation, urbanization, and population growth) have been examined in an attempt to elucidate the basis for this incessant drive for the maximal utilization of the bounties of Nature over world history.

All the previous chapters have delineated that accumulation, urbanization, and population growth within a world system as early as five thousand years ago circumscribed the overall dynamic of the evolution of human communities through world history. This overall trajectory is, of course, shaped and perhaps directed by hierarchical differentiation of the social structures, and the consequent power relations (Mann 1986, 1993).

As world history has marched on, despite the appearance of environmental movements of protest, I suspect that the world historical process of incessant accumulation via materialistic acquisition and consumption has advanced the process of the denaturalization of the human community in terms of identity orientation and formation. Per contra to Bergesen's stages of alienation (1995), not only is the human being alienated from his/her social being, pari passu s/he is also alienated from his/her natural beingness. Such a denaturalization process increases the distantiation from Nature in which the human is a part. Coupled with geographic separation from the wilds as a consequence of urbanization, humans are further denaturalized. Like Enkidu in the epic of Gilgamesh, who was enticed to leave the forest for urban life, the culturalization process of human communities has spiraled upwards unabated with the urbanization process. Culturalization de-naturalizes the human community.

Nature as a whole, which forms the basis of the materialistic reproduction of human communities, has thus been severely degraded as a consequence of these world historical processes. Ecological crisis conditions have ensued and, in certain cases, fostered the collapse of civilizations and empires, as discussed in chapters 2 and 3. Such periodic ecological crises, which seem commensurate with long economic downturns and might have also triggered them, suggest that in the end instead of "economy in command" it has always been "ecology in command" in the last instance, as the previous eight chapters of this book have indicated. It is only anthropocentrism which leads us to think otherwise. On this basis, Frank (1993), who has acknowledged the importance of ecological conditions in understanding the dynamics of historical transformations over the last five thousand years of world history, continues to refrain from doing this, but instead focuses on the overall accumulation process (Frank 1998).

If it is "ecology in command" that is the key for our understanding of system transformations, then perhaps periods of systemic crisis that are socioeconomic in character might have roots that are ecological in character. In this respect, the periods we know of in world history of prolonged socioeconomic and cultural decline, commonly referred to as the "dark ages," might have their origins in the severe degradation of the environment by the respective cultures, as we have shown in chapters 2–5.

Still, despite these ecological crises, the predominant orientation continues to be the maximal utilization of resources for the most gains, even with social-resistance movements emerging throughout world history during various ages to highlight the degradative impact of human activity on the environment, and, in some cases, pursuing alternative lifestyles and consumptive habits (Chew 1995a, 1995b). Since over the last five thousand years human communities have demonstrated a propensity for collective learning processes, one would expect that these communities would have realized the historical consequences of degradation of the environment from the fatal ecological history of the degradative encounters of other human communities. Unfortunately, this does not seem to have been the case even today. Perhaps Jurgen Habermas is right in his assessment of human social evolution (Habermas 1975). He has conjectured that the fundamental mechanism for social evolution in general is an automatic "inability not to learn. . . . Therein lies, if you will, the rationality of man. Only against this background does the overpowering irrationality of the history of the species become visible" (1975, 15). To this extent, for all human civilizations, it is not the problem of learning, it is *not-learning* that is the problem. Having stated this, how do we account for the relation between Culture (human communities) and the other parts of Nature over the long term? Are we natural aliens who have discovered a niche in the ecosystem, and have flourished socioeconomically and exploded in terms of numbers without any predatory checks, as Neil Evernden (1985) puts it? Our historical examination in this book of the Culture–Nature relations through the last five thousand years of human history would tend to agree with such a theme. Notwithstanding this, human occupation of the ecosystem can also be characterized as maladaptations that are increasingly showing signs of malignancy (Gregg 1955; Hern 1993, 1994). Hern (1993, 1994), in distinguishing the nature of human communities within the global ecosystem, has argued that it is similar to a malignant cellular process: rapid, uncontrolled growth; invasion and destruction of adjacent tissues; de-differentiation (loss of specialized appearance, morphology, and function at individual cell level); and metastasis (distant colonization). Paralleling this in the case of human communities as a whole are rapid, uncontrolled growth of global population; invasion and destruction of all planetary ecosystems, such as with deforestation, pollution, etc.; colonization and urbanization, including settlement fission; and increasing undifferentiation in structure and appearance as a consequence of the homogenization of cultural forms via the dynamics of the world economy. In medical diagnoses, according to Hern (1993, 771), confirmation of any two characteristics in the above cellular-differentiation process would establish the conditions for malignancy. It would be safe to say that from our examination of ecological relations throughout the last five thousand years of world history, these criteria have often been met by the human communities across time and space.

NOTES

An earlier version of this chapter was presented at the International Studies Association Annual Meeting, Theme Panel: Mentalities, World Views, and World Systems. Washington, D.C., February 18, 1999. Many thanks to Stefan Mosemann for his research on literature sources for this chapter.

1. This translation from *The Vedic Experience,* edited and translated by Raimundo Panikkar. Copyright © 1977 by University of California Press. Reprinted by permission of Raimundo Panikkar.

2. "Bearing witness" means to stand forth or to be an example for others. According to Devall (1992, 59), this means that it involves "the exploration of their [the participants'] own spiritual awakening in an age of environmental crisis, this means bearing witness for the mute forests that are being burned and chainsawed in the name of progress, greed, or ignorance."

3. For Devall (1991, 138), this means action to resist those interests that exploit the earth for "narrow human goals" and to "obey the laws of ecology."

Bibliography

Abu-Lughod, Janet. 1989. *Before European Hegemony*. New York: Oxford University Press.

———. 1993. "Discontinuities and Persistence: One World System or a Succession of Systems." In *The World System: Five Hundred Years or Five Thousand*, edited by A. G. F. Frank and B. K. Gills. London: Routledge.

Adams, Madonna. 1997. "Environmental Ethics in Plato's *Timaeus*." In *The Greeks and the Environment*, edited by L. Westra and T. Robinson. London: Rowman and Littlefield.

Adams, Richard E. W. 1983. "Ancient Land Use and Culture History in the Passion River Region." In *Prehistoric Settlement Patterns*, edited by R. E. W. Adams. Albuquerque: University of New Mexico Press.

Adams, Robert Mc C. 1981. *Heartland of the Cities*. Chicago: University of Chicago Press.

Adas, Michael. 1983. "Colonization, Commercial Agriculture and Destruction of the Deltaic Rainforests of British Burma in the Late Nineteenth Century." In *Global Deforestation and the Nineteenth-Century World Economy*, edited by R. Tucker and J. Richards. Durham: Duke University Press.

Adorno, Theodor. 1975. *The Positivist Dispute in German Sociology*. London: MacMillan.

Adorno, Theodor, and Max Horkheimer. 1979. *The Dialectic of Enlightenment*. New York: Seabury.

Adshead, S. A. M. 1974. "An Energy Crisis in Early Modern China." *Ching Shih Wen-t'i* 2, no. 2:20–28.

———. 1979. "Timber as a Factor in Chinese History: Problems, Sources, and Hypotheses." *Proceedings of the First International Symposium on Asian Studies*. Vol. 1. Hong Kong: Hong Kong University Press.

Agrawal, D. P. 1971. *The Copper Bronze Age in India*. New Delhi: Munshiram Manoharlal.

Agrawal, D. P., and R. K. Sood. 1982. "Ecological Factors and the Harappan Civilization." In *Harappan Civilization: A Contemporary Perspective*, edited by G. Possehl. New Delhi: Oxford University Press.

Albion, R.G. 1926. *Forests and Seapower: The Timber Problem of the Royal Navy*. Cambridge: Harvard University Press.

Algaze, Guillermo. 1989. "The Uruk Expansion: Cross-Cultural Exchange in Early Mesopotamia." *Current Anthropology* 30, no. 5:571–608.

———. 1993a. *The Uruk World System*. Chicago: University of Chicago Press.

———. 1993b. "Expansionary Dynamics of Some Early Pristine States." *American Anthropologist* 95 no. 2:304–33.

Allchin, Bridget. 1982. *The Rise of Civilization in India and Pakistan*. New York: Cambridge University Press.

Allen, Mitchell. 1992. "The Mechanisms of Underdevelopment: An Ancient Mesopotamian Example." *Review* 15, no. 3:453–76.

Amin, Samir. 1974. *Accumulation on the World Scale*. New York: Monthly Review Press.

———. 1994. "The Future of Global Polarization." *Review* 17, no. 3:337–47.

Amin, Samir, Giovanni Arrighi, Andre Gunder Frank, Immanuel Wallerstein. 1990. *Transforming the Revolution*. New York: Monthly Review Press.

Anderson, C. H. 1976. *The Sociology of Survival*. Homewood, Ill.: Dorsey.

Anderson, Perry. 1974. *Passages from Antiquity to Feudalism*. London: New Left.

Arnold, David. 1996. *The Problem of Nature: Environment, Culture, and European Expansion*. Cambridge, Mass.: Blackwell.

Asthana, Shashi. 1982. "Harappan Trade in Metals and Minerals: A Regional Approach." In *Harrapan Civilization: A Contemporary Perspective*, edited by G. Possehl. New Delhi: Oxford University Press.

———. 1993. "Harappan Trade in Metals and Minerals." In *Harrapan Civilization: A Recent Perspective*, edited by Gregory Possehl. New Delhi: Oxford University Press.

Aston, T., and C. Philpin, eds. 1985. *The Brenner Debate*. Cambridge: Cambridge University Press.

Athanasiou, Tom. 1992. "After the Summit." *Socialist Review* 22, no. 4:57–92.

Attenborough, David. 1987. *The First Eden: The Mediterranean World and Man*. Boston: Little Brown and Company.

Bag, A. K. 1985. *Science and Civilization in India*. New Delhi: Navrang.

Bahro, Rudolf. 1982. *Socialism and Survival*. London: Heretic.

———. 1984. *From Red to Green*. London: Verso.

Balassa, B. 1982. *Development Strategies in the Semi-Industrial Countries*. Baltimore: Johns Hopkins University Press.

Balcer, Jack Martin. 1974. "The Mycenaen Dam at Tiryns." *American Journal of Archaeology* 78, no. 2:77–149.

Barbosa, Luiz. 1993. "The World-System and the Destruction of the Brazilian Amazon Rainforest." *Review* 16, no. 2:215–40.

Bard, Kathryn. 1987. "Geography of Excavated Predynastic Sites and the Rise of Complex Society." *Journal of American Research Center in Egypt* 24:81–93.

Bari, Judi. 1993. "Red Green Politics." *Capitalism, Nature, Socialism* 4, no. 4:1–30.

Barkas, Janet. 1975. *The Vegetable Passion*. New York: Charles Scribner's Sons.

Barnet, D. L., and J. Cavanagh. 1996. *Global Dreams: Imperial Corporations and the New World Order*. New York: Simon and Schuster.

Bartley, T., and Albert Bergesen. 1997. "World-System Studies of the Environment." *Journal of World System Studies* 3, no. 3:369–80.

Barton, Gregory. 1999. "Enlightenment Foundations of World Environmentalism: Comparative Civilizations." *Review* 40:7–30.

Bauer, P. 1981. *Equality, the Third World and Economic Delusion*. London: Methuen.

———. 1984. *Reality and Rhetoric: Studies in the Economics of Development*. London: Weidenfeld and Nicholson.

Bautista, G. 1990. "The Forestry Crisis in the Philippines: Nature, Causes and Issues." *The Developing Economies* 27, no. 1:67–94.

Bayard, D. T. 1979. "The Chronology of Prehistoric Metallurgy in North-east Thailand." In *Early Southeast Asia*, edited by R. B. Smith and W. Watson. Kuala Lumpur: Oxford University Press.

———. 1980. "The Roots of Indochinese Civilization: Recent Developments in the Prehistory of Southeast Asia." *Pacific Affairs* 53, no. 1:89–114.

———. 1992. "Models, Scenarios, Variables, and Supposition: Approaches to the Rise of Social Complexity in Mainland Southeast Asia 700 B.C.–A.D. 500." In *Early Metallurgy, Trade, and Urban Centres in Thailand and Southeast Asia*, edited by D. T. Bayard. Bangkok: White Lotus.

Begley, Vimala, and Richard Daniel De Puma, eds. 1991. *Rome and India: The Ancient Sea Trade*. Madison: University of Wisconsin Press.

Belcher, M., and A. Gennino, eds. 1993. *Southeast Asia Rainforests*. San Francisco: Rainforest Action Network.

Bell, Martin, and John Boardman, eds. 1992. *Past and Present Soil Erosion*. Oxford: Oxbow Books.

Bello, W. 1982. *Development Debacle: The World Bank in the Philippines*. Berkeley: University of California Press.

Bellwood, Peter S. 1978. *Man's Conquest of the Pacific: The Prehistory of Southeast Asia and Oceania*. Auckland: Collins.

———. 1985. *Prehistory of Indo-Malayan Archipelago*. Sydney: Academic Press.

Bennet, J. 1990. "Knossos in Context: Comparative Perspectives on Linear B Administration of LMII-II Crete." *American Journal of Archaeology* 94:193–211.

Bentley, Jerry H. 1996. "Cross Cultural Interaction and Periodization in World History." *American Historical Review* 101, no. 3:749–70.

Benton, Ted. 1989. "Marxism and Natural Limits: An Ecological Critique and Reconstruction." *New Left Review* 178:51–86.

———. 1992. "Ecology, Socialism, and the Mastery of Nature: A Reply to Reiner Grundmann." *New Left Review* 194:55–75.

———. 1996. "From World-Systems Perspective to Institutional World History: Culture and Economy in Global Theory." *Journal of World History* 7, no. 2:201–95.

Bergesen, Albert. 1982. *Crisis in the World System*. Beverly Hills: Sage Publications.

———. 1995. "Eco-Alienation." *Humboldt Journal of Social Relations* 21, no. 1:111–26.

———. 1996. "Let's Be Frank about World History." In *Civilizations and World Systems*, edited by Stephen Sanderson. Walnut Creek, Cal.: Altamira Press.

Betancourt, Philip. 1976. "The End of the Greek Bronze Age." *Antiquity* 50, no. 197:40–47.

Bibby, T. G. 1970. *Looking for Dilmun*. London: Methuen.

Biehl, Janet, and Peter Staudenmaier. 1995. *Ecofascism Lessons from the German Experience*. San Francisco: A. K. Press.

Bilimoria, P. 1998. "Indian Religious Traditions." In *Spirit of the Environment: Religion, Value and Environmental Concern*, edited by David Cooper and J. Palmer. New York: Routledge.

Billamboz, Andre. 1990. *Siedlungsarchaologie im Alpenvorland II*. Stuggart: Konrad Theiss.

Bilsky, L. 1980. "Ecological Crisis and Response in Ancient China." In *Historical Ecology*, edited by L. Bilsky. New York: Kennikat Press.

Bintliff, J. 1982. "Climate Change, Archaeology and Quarternary Science in the Eastern Mediterranean Region." In *Climate Change in Later Prehistory*, edited by A. F. Harding. Edinburgh: Edinburgh University Press.

———. 1992. "Erosion in the Mediterranean Lands: A Reconsideration of Pattern, Process, and Methodology." In *Past and Present Soil Erosion*, edited by M. Bell and J. Boardman. Oxford: Oxbow.

Birch, Charles, and John Cobb. 1981. *The Liberation of Life: From the Cell to the Community*. Cambridge: Cambridge University Press.

Bird, Isabella L. 1967. *The Golden Chersonese and the Way Thither*. Kuala Lumpur: Oxford University Press.

Blaut, J. M. 1992. *1492: The Debate on Colonialism, Eurocentrism, and History*. Trenton: University of Pennsylvania Press.

———. 1993. *The Colonizer's Model of the World*. New York: Guilford Press.

Blazquez, J. M. 1992. "The Latest Work on the Export of Baetican Olive Oil to Rome and the Army." *Greece and Rome* 39, no. 2:173–88.

Bloom, Irene. 1997. "Human Nature and Biological Nature in Mencius Philosophy." *East and West* 47, no. 1:23–37.

Bookchin, Murray. 1982a. "Finding the Subject: Notes on Whitebook and Habermas." *Telos* 52:56–78.

———. 1982b. *The Ecology of Freedom*. Palo Alto: Chesire.

———. 1989. *Remaking Society: Pathways to a Green Future*. Boston: Southend Press.

———. 1990. *The Philosophy of Social Ecology: Essays on Dialectical Naturalism*. New York: Black Rose.

———. 1991. "Racism and the Future of the Movement." In *Defending the Earth*, edited by Steve Chase. Boston: Southend Press.

Boomgaard, Peter. 1988. "Forests and Forestry in Colonial Java." In *Changing Tropical Forests*, edited by J. Dargavel. Canberra, Aus.: Centre for Resource and Environment Studies.

———. 1998. "The VOC Trade in the Seventeenth Century." In *Nature and the Orient*, edited by Richard Grove. New Delhi: Oxford University Press.

Booth, David. 1985. "Marxism and Development Sociology: Interpreting the Impasse." *World Development* 13, no. 7:761–87.

Bosworth, Andrew. 1995. "World Cities and World Economic Cycles." In *Civilizations and World Systems*, edited by Stephen Sanderson. Walnut Creek, Cal.: AltaMira Press.

Bramwell, Anna. 1989. *Ecology in the Twentieth Century*. New Haven: Yale University Press.

Branigan, K. 1981. "Minoan Civilization." *Annual of the British School at Athens* 76:23–33.

———. 1984. "Minoan Community Colonies in the Aegean." In *The Minoan Thalassocracy: Myth and Reality*, edited by Robin Haag and Nanno Marinatos. Stockholm: Paul Astroms Forlag.

———. 1989. "Minoan Foreign Relations in Transition." *Aegaeum* 3:65–71.

Braudel, Fernand. 1972. *The Mediterranean and the Mediterranean World in the Age of Philip II*. Vol. 1. London: Fontana.

———. 1981. *The Structure of Everyday Life*. Vol. 1. New York: Harper and Row.

———. 1982. *The Wheels of Commerce*. Vol. 2. New York: Harper and Row.

———. 1984. *The Perspective of the World*. Vol. 3. New York: Harper and Row.

———. 1989. *The Identity of France*. Vols. 1–2. New York: Harper and Row.

Brinkman, J. A. 1968. "Ur: The Kassite Period and the Period of Assyrian Kings." *Orientalia* 38:310–48.

Brown, Neville. 1994. "Climate Change and Human History: Some Indicators from Europe A.D. 400–1400." *Environmental Review* 83, no. 1–2:37–43.

Brulle, Robert J. 1996. "Environmental Discourse and Social Movement Organization." *Sociological Inquiry* 66:58–63.

Brundtland Commission. 1987. *Our Common Future*. Oxford: Oxford University Press.

Bryant, Raymond. 1996. *The Political Economy of Forestry in Burma 1826–1993*. Honolulu, Hi.: University of Hawaii Press.

Bryson, R. A., and C. Padoch. 1980. "On the Climates of History." *Journal of Interdisciplinary History* 10, no. 4:583–97.

Bryson, R., and T. Swain. 1981. "Holocene Variations of Monsoon Rainfall in Rajasthan." *Quarternary Research* 16:135–45.

Bryson, R. A., H. H. Lamb, and David L. Donley. 1974. "Drought and the Decline of Mycenae." *Antiquity* 48, no. 189:46–50.

Bunker, Stephen. 1985. *Underdeveloping the Amazon: Extraction, Unequal Exchange, and the Failure of the Modern State*. Urbana, Ill.: University of Illinois Press.

Burawoy, Michael, and Janos Lukacs. 1992. *The Radiant Past*. Chicago: University of Chicago Press.

Burch, W. R. 1971. *Daydreams and Nightmares: A Sociological Essay on American Environment*. New York: Harper and Row.

Bures, R. 1991. "Ethical Dimensions of Human Attitude to Nature." *Radical Philosophy* 57:34–56.

Burns, Thomas J., Edward L. Kick, David A. Murray, and Dixie A. Murray. 1994. "Demography, Development, and Deforestation in a World-System Perspective." *International Journal of Comparative Sociology* 35:221–39.

Buttel, F., and P. Taylor. 1994. "Environmental Sociology and Global Environmental Change." In *Social Theory and the Global Environment*, edited by M. Redclift and T. Benton. London: Routledge.

Butti, Ken, and John Perlin. 1980. *A Golden Thread: 2,500 Years of Solar Architecture and Technology*. Palo Alto, Cal.: Chesire.

Butzer, Karl W. 1971. *Environment and Archaeology*. Chicago: Aldine and Atherton.

Byron, R., and M. A. Quintos. 1988. "Log Export Restrictions and Forest Industries Development in Southeast Asia 1975–1986." In *Changing Tropical Forests*, edited by J. Dargavel. Canberra, Aus.: Centre for Resource and Environmental Studies.

Cadogan, G. 1983. "Early Minoan and Middle Minoan Chronology." *American Journal of Archaeology* 87:507–18.

———. 1984. "A Minoan Thalassocracy." In *The Minoan Thalassocracy: Myth or Reality*, edited by Robin Hagg and Nanno Marinatos. Stockholm: Paul Astroms Forlag.

Callicott, Baird. 1986. "The Metaphysical Implication of Ecology." *Environmental Ethics* 8:301–16.

———. 1994. *Earth's Insights: A Multicultural Survey of Ecological Ethics from the Mediterranean Basin to the Australian Outback*. Berkeley, Cal.: University of California Press.

Carpenter, Rhys. 1966. *Discontinuity in Greek Civilization*. Cambridge: Cambridge University Press.

Carson, Rachel. 1962. *Silent Spring*. New York: Houghton Mifflin.

Carswell, John. 1991. "The Part of Mantai, Sri Lanka." In *Rome and India*, edited by V. Begley and D. Puma. Madison: University of Wisconsin Press.

Carter, Vernon, and Tom Dale. 1974. *Topsoil and Civilization*. Norman: University of Oklahoma Press.

Casson, Lionel. 1954. "Trade in the Ancient World." *Scientific American* 191, no. 5:98–104.

Catton, William. 1980. *Overshoot: The Ecological Basis for Revolutionary Change*. Urbana: University of Illinois Press.

Catton, William, and R. Dunlap. 1978. "Environmental Sociology: A New Paradigm." *American Sociologist* 13:41–49.

———. 1980. "A New Ecological Paradigm for Post-Exhuberant Sociology." *American Behavioral Scientist* 24:15–47.

Chadwick, J. 1972. "Life in Mycenaean Greece." *Scientific American* 227, no. 4:36–44.

———. 1976. *The Mycenaean World*. New York: Cambridge University Press.

Chakrabarti, D. K. 1990. *The External Trade of the Indus Civilization*. New Delhi: Munshiram Manoharlal Publications.

———. 1993. "Long-Barrel Cylinder Beads and the Issue of Pre-Sargonic Contact between Harappan Civilization and Mesopotamic." In *Harappan Civilization: A Recent Perspective*, edited by Gregory Possehl. New Delhi: Oxford University Press.

Chan, Wing-Tsit. 1963. *A Source Book in Chinese Philosophy*. Princeton: Princeton University Press.

Chandler, T. 1974. *Three Thousand Years of Urban Growth*. New York: Academic Press.

———. 1987. *Four Thousand Years of Urban Growth*. Lewiston, N.J.: Edwin Mellen Press.

Chao, Kang. 1986. *Man and Land in Chinese History*. Stanford, Cal.: Stanford University Press.

Chapple, Christopher. 1993a. *Nonviolence to Animals: Earth and Self in Asian Traditions*. Albany: State University of New York Press.

Chapple, Chrisotopher, ed. 1993b. *Worldviews and Ecology*. Lewiston, Penn.: Bucknell University Press.

Chase, Steve, ed. 1991. *Defending the Earth*. Boston: Southend Press.

Chase-Dunn, Christopher, ed. 1982. *Socialist States in the World-System*. Beverly Hills: Sage Publications.

———. 1989. *Global Formation*. Boston: Blackwell.

Chase-Dunn, Christopher, and T. Hall. 1991. "Conceptualizing Core–Periphery Hierarchies for Comparative Studies." In *Core–Periphery Relations in Precapitalist Worlds*, edited by C. Chase-Dunn and T. Hall. Boulder: Westview Press.

———. 1997a. *Rise and Demise: Comparing World-Systems*. Boulder: Westview Press.

———. 1997b. "Ecological Degradation and the Evolution of World Systems." *Journal of World-Systems Research* 3, no. 3:403–31.

Chaudhuri, K. N. 1978. *The Trading World of Asia and the East India Company 1660–1760*. Cambridge: Cambridge University Press.

———. 1985. *Trade and Civilization in the Indian Ocean: An Economic History from the Rise of Islam to 1750*. Cambridge: Cambridge University Press.

Cheney, Jim. 1987. "Ecofeminism and Deep Ecology." *Environmental Ethics* 27:115–45.

Cherry, J. F. 1986. "Polities and Palaces: Some Problems in Minoan State Formation." In *Peer Polity Interaction and Socio-Political Change*, edited by Colin Renfrew and J. F. Cherry. Cambridge: Cambridge University Press.

Chew, Sing C. 1992. *Logs for Capital: The Timber Industry and Capitalist Enterprise in the Nineteenth Century*. Westport, Conn.: Greenwood Press.

———. 1995a. "Environmental Transformations: Accumulation, Ecological Crisis, and Social Movements." In *A New World Order?: Global Transformation in the Late Twentieth Century*, edited by Jozsef Borocz and David A. Smith. Westport, Conn.: Praeger.

———. 1995b. "Environmental Imperatives and Development Strategies: Challenges for Southeast Asia." In *Asia: Who Pays for Growth?*, edited by Jayant Lele. London: Dartmouth.

———. 1997a. "Accumulation, Deforestation, and World Ecological Degradation 2500 B.C. to A.D. 1990." In *Advances in Human Ecology*, vol. 6, edited by Lee Freese. Westport, Conn.: JAI Press.

———. 1997b. "For Nature: Deep Greening World-Systems Analysis for the Twenty-First Century." *Journal of World-Systems Research* 3, no. 3:381–402.

———. 1999. "Ecological Relations and the Decline of Civilizations in the Bronze-Age World System: Mesopotamia and Harappa 2500 B.C.–1700 B.C." In *Environment and the World System*, edited by W. Goldfrank, David Goodman, and A. Szasz. Greenwich, Conn.: Greenwood Press.

Chew, Sing C., and Denemark, Robert K. 1996. "On Development and Underdevelopment." In *The Underdevelopment of Development*, edited by Sing C. Chew and Robert Denemark. Thousand Oaks, Cal.: Sage Publications.

Childe, Gordon. 1942. *What Happened in History*. Harmondsworth, Eng.: Penguin.

———. 1950. "The Urban Revolution." *Town Planning Review* 21:3–17.

———. 1957. "The Bronze Age." *Past and Present* 12:2–15.

———. 1964. *What Happened in History*. Harmondsworth, Eng.: Penguin.

Christensen, Peter. 1993. *The Decline of Iranshahr*. Copenhagen: Museum Tusculanum Press.

Christie, A. H. 1979. "Lin-I, Fu-nan, Java." In *Early Southeast Asia*, edited by R. B. Smith and W. Weston. Kuala Lumpur: Oxford University Press.

Christie, Jan W. 1990. "Trade and State Formation in the Malay Peninsula and Sumatra 300 B.C.–A.D. 700." In *The Southeast Asian Port and Polity*, edited by J. Kathirithamby-Wells and John Villiers. Singapore: Singapore University Press.

Chu, K. C. 1926. "Climate Pulsations During Historic Times in China." *Geographical Review* 16:274–81.

Cippola, C. M. 1976. *Before the Industrial Revolution: European Society and Economy 1000–1700*. New York: Norton.

Clark, A. H. 1949. *The Invasion of New Zealand by People, Plants, and Animals: The South Island*. New Brunswick, N.J.: Rutgers University Press.

Clark, Mary. 1989. *Ariadne's Thread: The Search for New Modes of Thinking*. New York: St. Martin's Press.

Clary, David. 1986. *Timber and the Forest Service*. Lawrence, Kan.: University of Kansas Press.

Coedes, George. 1983. *The Making of Southeast Asia*. Berkeley, Cal.: University of California Press.

Colchester, Marcus. 1994. "The New Sultans: Asian Loggers Move in on Guyana's Forest." *The Ecologist* 24, no. 1:45–52.

Coleman, James S. 1990. *The Foundations of Social Theory*. Cambridge: Cambridge University Press.

Colinvaux, P. A. 1980. *The Fates of Nations: A Biological Theory of History*. London: Harmondsworth.

Colless, B. E. 1969. "Persian Merchants and Missionaries in Medieval Malaya." *Journal of the Malayan Branch of the Royal Asiatic Society* 42, no. 2:10–47.

Cook, Earleen H. 1985. *Desertification and Deforestation*. Monticello, Ill.: Vance.

Cornell, T. J. 1995. "Warfare and Urbanization in Roman Italy." In *Urban Society in Roman Italy*, edited by T. Cornell and K. Lomas. New York: St. Martin's Press.

Cotgrove, Stephen. 1982. *Catastrophe or Cornucopia: The Environment, Politics, and the Future*. Boston: Wiley.

Coulton, J. J. 1988. "Greek Building Techniques." In *Civilization of the Ancient Mediterranean*, edited by M. Grant and R. Kitzinger. 3 vols. New York: Charles Scribner and Sons.

Cox, Thomas R. 1985. "The Impact of Western Expansion on World Ecosystems." In *Environmental History*, edited by Kendall Bailes. New York: University Press of America.

Crawford, Harriet. 1973. "Mesopotamia's Invisible Exports in the Third Millennium B.C." *World Archaeology* 5, no. 2:232–41.

———. 1998. *Dilmun and Its Neighbors*. Cambridge: Cambridge University Press.

Cresmachi, Mauro, and Vincent C. Piggott. 1992. "Paleoenvironment and Late Prehistoric Sites in the Lopburi Region of Central Thailand." In *Southeast Asian Archaeology*, edited by Ian Glover. Hull: Centre for Southeast Asian Studies.

Crone, Patricia. 1987. *Meccan Trade and the Rise of Islam*. London: Basil Blackwell.

Cronon, William, ed. 1995. *Uncommon Ground*. New York: Norton.

Crosby, Alfred. 1972. *The Columbian Exchange: Biological and Cultural Consequences of 1492*. Westport, Conn.: Greenwood Press.

———. 1986. *Ecological Imperialism: The Biological Expansion of Europe 900–1900*. Cambridge: Cambridge University Press.

Curtin, Philip. 1984. *Cross-Cultural Trade in World History*. New York: Cambridge University Press.

Dales, G. F. 1977. "Shifting Trade Patterns between Iranian Plateau and the Indus Valley." In *Le Plateau Iranien et l'Asie Centrale*, edited by J. Deshayes. Paris: CNRS.

———. 1979. "The Decline of the Harappans." In *Ancient Cities of the Indus*, edited by G. Possehl. New Delhi: Vikas.

Dalfes, H. N., G. Kukla, and H. Weiss, eds. 1996. *Climate Change in the Third Millennium B.C.* Frankfurt: Springer.

Dalton, Russell. 1994. *The Green Rainbow: Environmental Groups in Western Europe*. New Haven: Yale University Press.

Daly, Herman. 1991. *Steady-State Economics*. Washington, D.C.: Island Press.

Daly, Herman, and John Cobb. 1989. *For the Common Good*. Boston: Beacon Press.

Dandokar, R. 1987. *Harappan Bibliography*. Poona: Bhandarkar Oriental Research Institute.

Dauvergne, Peter. 1997. *Shadows in the Forest: Japan and the Politics of Timber in Southeast Asia*. Cambridge: MIT Press.

Dean, W. 1983. "Deforestation in Southeastern Brazil." In *Global Deforestation in the Nineteenth-Century World Economy*, edited by R. P. Tucker and J. Richards. Durham, N.C.: Duke Press Policy Studies.

DeBary, Wm. Theodore. 1958. *Sources of Indian Tradition*. New York: Columbia University Press.

Demeny, Paul. 1990. "Population." In *The Earth as Transformed by Human Action*, edited by B. L. Turner, William C. Clark, Robert W. Kates, John Richards, Jessica Mathews, and William B. Meyer. Cambridge: Cambridge University Press.

Denemark, Robert, Jonathan Friedman, Barry K. Gills, and George Modelski, eds. 2000. *World System History: The Social Science of Long-Term Change*. London: Routledge.

Denevan, W. M. 1973. "Development and the Imminent Demise of the Amazon Rainforest." *Professional Geographer* 25:130–35.

Deo, S. B. 1991. "Roman Trade: Recent Archaeological Discoveries in Western India." In *Rome and India*, edited by V. Begley and R. De Puma. Madison: University of Wisconsin Press.

Desborough, V. R. 1972. *The Greek Dark Ages*. London: Ernest Benn.

———. 1975. " The End of Mycenaean Civilization and the Dark Age." In *Cambridge Ancient History*, edited by I. E. S. Edwards. Cambridge: Cambridge University Press.

Descartes, René. 1901. *The Method, Meditations, and Philosophy of Descartes*. New York: Tudor.

———. 1983. *Principles of Philosophy*. Boston: Reidel.

Detienne, Marcel, and Jean Pierre Vernant. 1989. *The Cuisine of Sacrifice among the Greeks*. Chicago: University of Chicago Press.

Devall, Bill. 1988. *Simple in Means, Rich in Ends*. Salt Lake City: Peregrine Smith.

———. 1991. "Political Activism in a Time of War." *ReVision* 13, no. 3:135–41.

———. 1992. "Deep Ecology and Radical Environmentalism." In *American Environmentalism*, edited by R. Dunlap and A. Mertig. Philadelphia: Taylor and Francis.

———. 1993 *Living Richly in an Age of Limits*. Salt Lake City: Peregrine Smith.

———. 1994. *Clear Cut: The Tragedy of Industrial Forestry*. San Francisco: Sierra Club.

Devall, Bill, and G. Sessions. 1987. *Deep Ecology*. Salt Lake City: Peregrine Smith.

Diakonoff, I. M. 1991a. "The City-States of Sumer." In *Early Antiquity*, edited by I. M. Diakonoff. Chicago: University of Chicago Press.

———. 1991b. "Early Despotisms in Mesopotamia." In *Early Antiquity*, edited by I. M. Diakonoff. Chicago: University of Chicago Press.

Diamond, Irene, and Gloria Orenstein. 1990. *Reweaving the World: The Emergence of Ecofeminism*. San Francisco: Sierra Club.

Dickens, Peter. 1992. *Society and Nature: Towards a Green Social Theory*. Philadelphia: Temple University Press.

Dickinson, O. T. 1977. *The Origins of Mycenaean Civilization*. Goteborg: SIMA.

———. 1994. *The Aegean Bronze Age*. Cambridge: Cambridge University Press.

DiZarga, Gus. 1995. "Deep Ecology and Liberal Modernity." *Social Theory and Practice* 21, no. 2:239–69.

Dobb, M. 1952. *Studies in the Development of Capitalism*. New York: International Publishers.

———. 1976. "A Reply." In *The Transition from Feudalism to Capitalism*, edited by R. Hinton. London: New Left Books.

Dornberg, John. 1992. "Battle of the Teutoberg Forest." *Archaeology* 45, no. 5:26–33.

Drengson, Alan. 1989. *Beyond Environmental Crisis: From Technocrat to Planetary Person*. New York: Peter Lang.

Drews, R. 1993. *The End of the Bronze Age*. Princeton: Princeton University Press.

Dubos, Rene. 1974. *Beast or Angel?* New York: Scribners.

Dumayne, Lisa. 1992. "Late Holocene Paleoecology and Human Impact on the Environment of Northern Britain." Ph.D. diss., University of Southampton.

Dumont, Louis. 1970. "Religion, Politics and Society in the Individualistic Universe." *Proceedings of the Royal Anthropological Institute of Great Britain and Ireland* 5:134–56.

Duncan-Jones, R. P. 1974. *The Economy of the Roman Empire*. Cambridge: Cambridge University Press.

———. 1990. *Structure and Scale in the Roman Economy*. Cambridge: Cambridge University Press.

Dunlap, R. E., and W. R. Catton. 1979. "Environmental Sociology." *Annual Review of Sociology* 5:243–73.

———. 1992–93. "Toward an Ecological Sociology: The Development, Current Status, and Probable Future of Environmental Sociology." *Annals of the International Institute of Sociology* 3:263–84.

———. 1994. "Struggling with Human Exemptionalism: The Rise, Decline, and Revitalization of Environmental Sociology." *American Sociologist* 25, no. 1:5–30.

Dunlap, Riley, et al. 1992. *The Health of the Planet Survey (Preliminary Report)*. Princeton: George H. Gallup International Institute.

Dunlap, Riley, and Angela Mertig, eds. 1992. *American Environmentalism*. Philadelphia: Taylor and Francis.

Dunn, F. L. 1975. "Rainforest Collectors and Traders: A Study of Resource Utilization in Modern and Ancient Malaya." Monographs of the Malayan Branch of the Royal Asiatic Society, no. 5. Kuala Lumpur: MBRAS.

Durand, John D. 1960. "The Population Statistics of China A.D. 2–1953." *Population Studies* 13, no. 2:209–56.

Dutta, Pratap C. 1983. *The Bronze Age Harappans*. Calcutta: Anthropological Survey of India.

Dyson, Robert. 1982. "Paradigm Changes in the Study of the Indus Civilization." In *Harappan Civilization: A Contemporary Perspective*, edited by G. Possehl. New Delhi: Oxford University Press.

Eckersley, Robyn. 1989. "Green Politics and the New Class." *Political Studies* 37:205–23.

———. 1992. *Environmentalism and Political Theory*. New York: SUNY Press.

Economic and Social Commission for Asia and the Pacific (ESCAP). 1990. *The State of the Environment for Asia and the Pacific*. Bangkok: ESCAP.

Edens, C. 1992. "The Dynamics of Trade in the Ancient Mesopotamian World-System." *American Anthropologist* 94:118–39.

———. 1993. "Indus and Arabian Interaction During the Bronze Age." In *Harappan Civilization: A Recent Perspective*, edited by G. Possehl. New Delhi: Oxford University Press.

Eder, Klaus. 1990. "The Cultural Code of Modernity and the Problem of Nature: A Critique of the Naturalistic Notion of Progress." In *Rethinking Progress: Movements, Forces, and Ideas at the End of the 20th Century*, edited by J. Alexander and P. Sztompka. New York: Unwin and Hyman.

———. 1993. *The New Politics of Class*. Thousand Oaks: Sage Publications.

———. 1996. *The Social Construction of Nature*. London: Sage Publications.

Edmonds, Richard Lomis. 1994. *Patterns of China's Lost Harmony: A Survey of the Country's Environmental Degradation and Protection.* London: Routledge.

Egan, Timothy. 1992. "Citing Space Photos, Scientists Say Forests in Northwest Are in Danger." *New York Times*, June 11, A13.

Ekholm, Kajsa. 1980. "On the Limits of Civilizations: The Structure and Dynamics of Global Systems." *Dialectical Anthropology* 5:155–66.

Ekholm, K., and J. Friedman. 1982. "'Capital' Imperialism and Exploitation in the Ancient World-Systems." *Review* 4, 1:87–109.

Ellis, Maria. 1976. *Agriculture and the State in Ancient Mesopotamia.* Philadelphia: Babylonian Fund.

Elvin, Mark. 1973. *The Patterns of China's Past.* Stanford, Cal.: Stanford University Press.

———. 1993. "Three Thousand Years of Unsustainable Growth: China's Environment from Archaic Times to the Present." *East Asian History* 6:7–46.

Elvin, Mark, and Tsui-Jung Liu, eds. 1998. *Sediments of Time: Environment and Society in Chinese History.* Cambridge: Cambridge University Press.

Emmanuel, Arghiri. 1972. *Unequal Exchange.* New York: Monthly Review Press.

Emsley, J. 1994. "Ancient World Was Poisoned." *New Scientist* 143, no. 1945.

Emzensberger, H. M. 1974. "Critique of Political Ecology." *New Left Review* 84:45–67.

Erdosy, George. 1998. "Deforestation in Pre- and Protohistoric South Asia." In *Nature and the Orient*, edited by R. Grove. New Delhi: Oxford University Press.

Evans, Arthur. 1921. *The Palace of Minos.* London: MacMillan.

Evernden, Neil. 1985. *The Natural Alien.* Toronto: University of Toronto Press.

———. 1993. *The Social Creation of Nature.* Baltimore: Johns Hopkins University Press.

Fairservis, Walter. 1979a. "The Harappan Civilization: New Evidence and More Theory." In *Ancient Cities of the Indus*, edited by G. Possehl. New Delhi: Vikas.

———. 1979b. "The Origin, Character, and Decline of an Early Civilization." In *Ancient Cities of the Indus*, edited by G. Possehl. New Delhi: Vikas.

Fang, Jin-Qi. 1992. "Establishment of a Data Bank from Records of Climatic Disasters and Anomalies in Ancient Chinese Documents." *International Journal of Climatology* 12:499–519.

Farber, Howard. 1978. "A Price and Wage Study for Northern Babylonia During the Old Babylonian Period." *Journal of the Economic and Social History of the Orient* 21:1–51.

Fattovich, Rodlfo. 1985. "Remarks on the Dynamics of State Formation in Ancient Egypt." In *On Social Evolution. Vienna Contributions to Ethnology and Anthropology*, edited by Walter Dostal. Vienna: Verlag Ferdinand Berger.

Fearnside, P. M. 1982. "Deforestation in the Brazilian Amazon: How Fast Is It Occurring?" *Ambro* 7:82–88.

Feuerwerker, A. 1990. "Chinese Economic History in Comparative Perspective." In *Heritage of China: Contemporary Perspective on Chinese Civilization*, edited by P. Ropp. Berkeley: University of California Press.

Finlay, Robert. 1992. "Portuguese and Chinese Maritime Imperialism: Camoes's Lusiads and Luo Maodeng's Voyage of the San Bao Eunuch." *Comparative Studies in Society and History* 34, no. 2:225–41.

Finley, M. I. 1973. *The Ancient Economy.* London: Penguin.

———. 1977. "The Ancient City: From Fustel de Coulanges to Max Weber and Beyond." *Comparative Studies in Society and History* 19, no. 3:305–27.

Flam, L. 1976. "Settlement Subsistence and Population: A Dynamic Approach to the Development of the Indus Valley Civilization." In *Ecological Backgrounds of South Asian Prehistory: Occasional Papers and Theses of the South Asian Program of Cornell University*, edited by K. Kennedy and G. Possehl. Ithaca: Cornell University.

Flannery, Kent V. 1965. "The Ecology of Early Food Production of Mesopotamia." *Science* 147:1247–56.

Fleming, Stuart. 1983. "Where Did All the Wood Go?" *Archaeology* 36, no. 4:66–67.

Forbes, R. J. 1964. *Studies in Ancient Technology*. Vol. 5. Leiden: Brill.

Foreman, Dave. 1992. *Confessions of an Eco-Warrior*. New York: Harmony Books.

Foreman, D., and H. Wolke. 1989. *The Big Outside*. New York: Harmony Books.

Foster, John. 1993. "The Limits of Environmentalism without Class: Lessons from the Ancient Forest Struggle of the Pacific Northwest." *Capitalism, Nature, Socialism* 4, no. 1:11–42.

———. 1994. *The Vulnerable Planet: A Short Economic History of the Environment*. New York: Monthly Review Press.

Foster-Carter, Aidan. 1991. "Development Sociology: Whither Now?" *Sociology Review* 1, no. 2:34–47.

Fox, S. 1985. *John Muir and His Legacy: The American Conservation Movement*. Madison: University of Wisconsin Press.

Fox, W. 1990. *Towards a Transpersonal Ecology: Developing New Foundations for Environmentalism*. Boston: Shambhala.

Francfort, H. P. 1986. "Archaeological and Environmental Researches in the Ghaggar (Saraswati) Plains." *Man and Environment* 10:97–100.

Frank, Andre Gunder. 1978. *World Accumulation*. New York: Monthly Review Press.

———. 1992. *The Centrality of Central Asia*. Amsterdam: University of Amsterdam Press.

———. 1993. "Bronze Age World System Cycles." *Current Anthropology* 34, no. 4:383–429.

———. 1995. "Transitional Ideological Modes: Feudalism, Capitalism, Socialism." *Critique of Anthropology* 11, no.2:171–88.

———. 1998. *ReOrient*. Berkeley, Cal.: University of California Press.

Frank, Andre Gunder, and B. K. Gills. 1992. "The Five-Thousand-Year World System: An Interdisciplinary Introduction." *Humboldt Journal of Social Relations* 18, no. 2:1–80.

Freeman, Charles. 1996. *Egypt, Greece, and Rome*. New York: Oxford University Press.

Freeman, Chris. 1984. *Long Waves in the World Economy*. London: Frances Pinter.

Friedman, J. 1992. "General Historical and Culturally Specific Properties of Global Systems." *Review* 15, no. 3:335–72.

Fritts, H. C. 1991. *Reconstructing Large-Scale Climatic Patterns from Tree-Ring Data*. Tucson: University of Arizona Press.

Frobel, Folker, Jurgen Heinrichs, and Otto Kreye. 1980. *The New International Division of Labor*. New York: Cambridge University Press.

Fukuyama, Francis. 1992. *The End of History*. New York: Free Press.

Fulford, Michael. 1992. "Territorial Expansion and the Roman Empire." *World Archaeology* 23, no. 3:294–305.

Gadd, C. J. 1932. "Seals of Ancient Indian Style Found at Ur." *Proceedings of the British Academy* 18:3–22.

———. 1979. "Seals of Ancient Indian Style Found at Ur." In *Ancient Cities of the Indus*, edited by G. Possehl. New Delhi: Vikas.

Gadgil, Madhav. 1988. "On the History of the Kannada Forests." In *Changing Tropical Forests*, edited by J. Dargavel. Canberra, Aus.: Centre for Resource and Environmental Studies.

Gadgil, Madhav, and Ramachandra Guha. 1992. *This Fissured Land: An Ecological History of India*. Berkeley: University of California Press.

Gale, Noel H., and Zofia Stos-Gale. 1981. "Lead and Silver in the Ancient Aegean." *Scientific American* 244, no. 6:176–92.

Gallant, Thomas. 1991. *Risk and Survival in Ancient Greece*. Cambridge: Cambridge University Press.

Garbsch, J. 1982. *Terra Sigillata: Ein Weltreich im Spiegel seines Luxusgeschirrs Prahistorische Staatssammlung*. Munich: Suhrkamp-Verlag.

Garnsey, Peter. 1988. *Famine and Food Supply in the Graeco-Roman World: Responses to Risk and Crisis*. Cambridge: Cambridge University Press.

Gedicks, Al. 1993. *The New Resource Wars*. Boston: Southend Press.

Gelb, I. J. 1973. "Prisoners of War in Early Mesopotamia." *Journal of Near Eastern Studies* 32:70–98.

Gereffi, Gary, and M. Korzniewicz. 1994. *Commodity Chains and Global Capitalism*. Westport, Conn.: Praeger.

Gibson, McGuire. 1970. "Violation of Fallow and Engineered Disaster in Mesopotamian Civilization." In *Irrigation's Impact on Society*, edited by T. Downing and M. Gibson. Tucson: University of Arizona Press.

———. 1973. "Population Shift and the Rise of Mesopotamian Civilization." In *The Explanation of Culture Change: Models in Prehistory*, edited by Colin Renfrew. London: Duckworth.

Gills, Barry K. 1996. "The 'Continuity Thesis' in World Development." In *The Underdevelopment of Development*, edited by Sing C. Chew and Robert K. Denemark. Thousand Oaks, Cal.: Sage Publications.

Gills, Barry K., and A. G. Frank. 1990. "The Cumulation of Accumulation: Theses and Research Agenda for Five Thousand Years of World System History." *Dialectical Anthropology* 15:19–42.

———. 1991. "Five Thousand Years of World System History: The Cumulation of Accumulation." In *Core–Periphery Relations in Precapitalist Worlds*, edited by C. Chase-Dunn and T. Hall. Boulder: Westview Press.

———. 1992. "World System Cycles, Crises, and Hegemonial Shifts 1700 B.C. to 1700 A.D." *Review* 15, no. 4:621–87.

Gimpel, Jean. 1976. "Environment and Pollution." In *The Medieval Machine: The Industrial Revolution of the Middle Ages*, edited by Jean Gimpel. New York: Holt, Rinehart and Winston.

Glacken, Clarence. 1967. *Traces of the Rhodian Shore*. Berkeley: University of California Press.

Glamann, K. 1958. *Dutch-Asiatic Trade 1620–1740*. Copenhagen: The Hague.

Glasser, Harold. 1995. "Toward a Descriptive Participatory Theory of Environmental Policy Analysis and Project Evaluation." Ph.D. diss., University of California, Davis.

Glotz, Gustave. 1926. *Ancient Greece at Work: An Economic History of Greece from the Homerica Period to Roman Conquest*. New York: Alfred A. Knopf.

Glover, Ian. 1979. "The Late Prehistoric Period in Indonesia." In *Early Southeast Asia*, edited by R. B. Smith and W. Watson. Oxford: Oxford University Press.

———. ed. 1992. *Early Metallurgy, Trade, and Urban Centers in Thailand and Southeast Asia*. Bangkok: White Lotus Press.

———. 1996. "The Archaeological Evidence for Early Trade between India and Southeast Asia." In *The Indian Ocean in Antiquity*, edited by Julian Reade. London: Kegan Paul.

Going, C. J. 1992. "Economic Long Waves in the Roman Period?" *Oxford Journal of Archaeology* 11, no. 1:93–117.

Goldfrank, W., David Goodman, and Andrew Szasz. 1999. *Ecology and the World-System*. Westport, Conn.: Greenwood Press.

Goldin, Owen. 1997. "The Ecology of Critias and Platonic Metaphysics." In *Environment and the Greeks*, edited by L. Westra and T. Robinson. London: Rowman and Littlefield.

Goldstein, Joshua. 1988. *Long Cycles: Prosperity and War in the Modern Age*. New Haven: Yale University Press.

Goldstone, Jack. 1985. "Ecological Dynamics of Empires: Seventeenth-Century Crises in Ottoman Turkey and Ming China." In *Comparative Social Dynamics*, edited by E. Cohen, M. Lissack, and Uri Almagor. Boulder: Westview Press.

———. 1991. *Revolution and Rebellion in the Early Modern World*. Berkeley: University of California Press.

Gomme, A. W. 1933. *The Population of Athens in the Fifth and Fourth Centuries B.C.* Chicago: Argonaut.

Gong, G., and S. Hameed. 1991. "The Variation of Moisture Conditions in China During the Last Two Thousand Years." *International Journal of Climatology* 11:271–83.

Goodland, R. J. A., and H. S. Irwin. 1975. *Amazon Jungle: Green Hell to Red Desert*. Amsterdam: Elsevier.

Gorny, Ronald L. 1989. "Environment, Archaeology, and History in Hittite Anatolia." *Biblical Archaeologist* 23:78–94.

Gorz, Andre. 1980. *Ecology as Politics*. Boston: Southend Press.

———. 1982. *Farewell to the Working Class*. London: Pluto.

Goto, Daisuke. 1993. "Logging in Siberia: Japan's Involvement." *Japan Environmental Exchange Newsletter*, July.

Gottlieb, Robert. 1993. *Forcing the Spring: The Transformation of the American Environmental Movement*. Washington, D.C.: Island Press.

Government of Malaysia. 1966. First Malaysia Plan 1966–1970. Kuala Lumpur.

———. 1981. Fourth Malaysia Plan 1981–1985. Kuala Lumpur.

———. 1986. Fifth Malaysia Plan 1986–1990. Kuala Lumpur.

Graham, Alistair. 1993. "Wood Flows around the Pacific Rim: A Corporate Picture." Paper presented at first International Temperate Forest Conference. Tasmania, Australia.

Greene, Kevin. 1986. *The Archaeology of the Roman Economy*. Berkeley: University of California Press.

Gregg, A. 1955. "A Medical Aspect of the Population Growth." *Science* 121:681–82.

Grey, William. 1986. "A Critique of Deep Ecology." *Journal of Applied Philosophy* 3:211–16.

Griffin, David. 1988. *The ReEnchantment of Science: Post Modern Proposals*. Albany, N.Y.: SUNY Press.

Griffiths, Tom, and L. Robin, eds. 1998. *Ecology and Empire: Environmental History of Settler Societies*. Seattle: University of Washington Press.

Grimal, Pierre. 1983. *Roman Cities*. Madison: University of Wisconsin Press.

Grove, Richard. 1989. "Scottish Missionaries, Evangelical Discourses and the Origins of Conservation Thinking in Southern Africa." *Journal of Southern African Studies* 15, no. 2:163–87.

———. 1990. "Colonial Conservation, Ecological Hegemony and Popular Resistance: Towards a Global Synthesis." In *Imperialism and the Natural World*, edited by John MacKenzie. Manchester: University of Manchester Press.

———. 1992. "Origins of Western Environmentalism." *Scientific American*, July, 42–47.

———. 1995. *Green Imperialism*. Cambridge: Cambridge University Press.

———. 1997. *Ecology, Climate, and Empire*. Cambridge: Cambridge University Press.

———. ed. 1998. *Nature and the Orient: The Environmental History of South and Southeast Asia*. New Delhi: Oxford University Press.

Grumbine, Edward. 1992. *Ghost Bears*. Covelo, Cal.: Island Press.

Grundmann, Reiner. 1991. "The Ecological Challenge to Marxism." *New Left Review* 187:24–42.

Guha, R. 1989. *The Unquiet Woods*. Berkeley: University of California Press.

———. 1992. "Prehistory of Indian Environmentalism." *Economic and Political Weekly* 27, no. 1:23–25.

Guha, R., and M. Gadgil. 1989. "State Forestry and Social Conflict in British India." *Past and Present* 123:141–77.

Ha, Van Tan. 1986. "Oc Eo: Endogenous and Exogenous Elements." *Vietnam Social Sciences* 1–2, no. 7–8:91–101.

Habermas, Jurgen. 1968. *Toward a Rational Society*. Boston: Beacon Press.

———. 1975. *Legitimation Crisis*. Boston: Beacon Press.

———. 1981, 1989. *The Theory of Communicative Action*. Vols. 1 and 2. Boston: Beacon Press.

Hagg, R., and N. Marinatos, eds. 1984. *The Minoan Thalassocracy: Myth and Reality*. Goteborg: P. Astrom Forlag.

Hall, K. 1982. "The Indianization of Funan: An Economic History of Southeast Asia's First State." *Journal of Southeast Asian Studies* 13:81–106.

———. 1985. *Maritime Trade and State Development in Early Southeast Asia*. Honolulu, Hi.: University of Hawaii Press.

———. 1992. "Economic History of Early Southeast Asia." In *Cambridge History of Southeast Asia*, vol. 1, edited by N. Tarling. Cambridge: Cambridge University Press.

Halstead, P. 1981. "From Determinism to Uncertainty: Social Storage and the Rise of the Minoan Palace." In *Economic Archaeology: Towards an Integration of Ecological and Social Approaches*, edited by A. Sheridan and G. N. Bailey. Oxford: British Archaeological Reports, no. 96.

———. 1994. "The North-South Divide: Regional Paths to Complexity in Prehistoric Greece." In *Development and Decline in the Mediterranean Bronze Age*, edited by Clay Mathes and Simon Stoddart. Sheffield: J. R. Collis Publications.

Hamblin, Robert L., and Brian L. Pitcher. 1980. "The Classic Maya Collapse: Testing Class Conflict Hypotheses." *American Antiquity* 45:246–67.

Hanson, W. S. 1988. "Administration, Urbanization, and Acculturation in the Roman West." In *The Administration of the Roman Empire*, edited by David C. Braund. Exeter: Short Run Press.

Hardin, Garrett. 1993. *Living within Limits: Ecology, Economies and Population Taboos*. New York: Oxford University Press.

Harding, A., ed. 1982. *Climatic Change in Later Prehistory*. Edinburgh: Edinburgh University Press.

——. 1984. *The Mycenaeans and Europe*. London: Academic Press.

Harris, Marvin. 1977. *Cannibals and Kings: The Origins of Culture*. New York: Random House.

——. 1979. *Cultural Materialism: The Struggle for a Science of Culture*. New York: Random House.

Harrison, Robert Pogue. 1992. *Forests: The Shadow of Civilization*. Chicago: University of Chicago Press.

Hartwell, Robert. 1967. "A Cycle of Economic Change in Imperial China: Coal and Iron in Northeast China A.D. 750–1350." *Journal of Economic and Social History* 10, no. 1:102–59.

——. 1982. "Demographic, Political and Social Transformations of China 750–1550." *Harvard Journal of Asiatic Studies* 42:365–442.

Harvey, David. 1996. *Justice, Nature, and the Geography of Difference*. Cambridge, Mass.: Blackwell Publishers.

Haselgrove, Colin. 1980. "Culture Process on the Periphery: Belgic Gaul and Rome During the Late Republic and Early Empire." In *Centre and Periphery in the Ancient World*, edited by Michael Rowlands, Mogens Larsen, and Kristian Kristiansen. Cambridge: Cambridge University Press.

Hecht, S., and Cockburn, A. 1990. *The Fate of the Forest*. New York: Harper.

Hedeager, Lotte. 1980. "Empire, Frontier, and the Barbarian Hinterland: Rome and Northern Europe from A.D. 1–400." In *Centre and Periphery in the Ancient World*, edited by Michael Rowlands, Mogens Larsen, and Kristian Kristiansen. Cambridge: Cambridge University Press.

Hern, Warren M. 1993. "Is Human Culture Carcinogenic for Uncontrolled Population Growth and Ecological Destruction?" *BioScience* 43, no. 11:768–73

——. 1994. "A Case of Malignant Maladaptation: Human Occupation of Global Systems." Paper presented at the 161st annual meeting of the American Association for the Advancement of Science. San Francisco, Cal.

Higham, Charles. 1988. *The Archaeology of Mainland Southeast Asia from 10,000 B.C. to the Fall of Angkor*. Cambridge: Cambridge University Press.

——. 1996. *The Bronze Age of Southeast Asia*. Cambridge: Cambridge University Press.

Hiller, Stefan. 1984. "Pax Minoica vs. Minoan Thalassocracy: Military Aspects of Minoan Culture." In *The Minoan Thalassocracy: Myth and Reality*, edited by Robin Hags and Nanno Marinatos. Stockholm: Paul Astrom Forlag.

Ho, Ping Ti. 1959. *Studies on the Population of China 1368–1958*. Cambridge: Harvard University Press.

Hobsbawm, Eric. 1968. *Industry and Empire*. London: Penguin.

Hodge, Trevor A. 1976. *Roman Aqueducts and Water Supply*. London: Gerald Duckworth.

Hohenberg, P., and L. Lees. 1985. *The Making of Urban Europe 1100–1950*. Cambridge: Cambridge University Press.

Hong, S., Jean-Pierre Candolone, Clair C. Patterson, Claude F. Boutron. 1994. "Greenland Ice Evidence of Hemispheric Lead Pollution Two Millennia Ago by Greeks and Romans." *Science* 265: 1841–43.

——. 1996. "History of Ancient Copper Smelting: Pollution During Roman and Medieval Times Recorded in Greenland Ice." *Science* 272:246–48.

Hood, M. S. F. 1983. "The Country House and Minoan Society." In *Minoan Society Proceedings of the Cambridge Colloquium*, edited by O. Krzyszkowska and L. Nixon. Bristol: Bristol Classical Press.

Hooker, J. T. 1976. *Mycenaean Greece*. London: Routledge.

Hopkins, Keith. 1978. "Economic Growth and Towns in Classical Antiquity." In *Towns in Societies*, edited by Peter Abrams and E. Wrigley. Cambridge: Cambridge University Press.

———. 1980. "Taxes and Trade in the Roman Empire 200 B.C.–A.D. 400." *Journal of Roman Studies* 70:101–25.

———. 1988. "Roman Trade, Industry, and Labor." In *Civilizations of the Ancient Mediterranean: Greece and Rome*, edited by Michael Grant and R. Kitzinger. New York: Charles Scribner and Sons.

Howgego, Christopher. 1992. "The Supply and Use of Money in the Roman World 200 B.C. to A.D. 300." *The Journal of Roman Studies* 82:1–31.

Hughes, Donald J. 1975. *Ecology in Ancient Civilizations*. Albuquerque: University of New Mexico Press.

———. 1982. "Deforestation, Erosion, and Forest Management in Ancient Greece and Rome." *Journal of Forest History* 26, no. 2:60–75.

———. 1983. "How the Ancients Viewed Deforestation." *Journal of Field Archaeology* 10:25–38.

———. 1994. *Pan's Travail*. Baltimore: Johns Hopkins University Press.

———. 1995. "Ecology and Development as Narrative Themes of World History." *Environmental History Review* 19:1–16.

———. 1997. "Mencius, Ecologist." *Capitalism, Socialism, Nature* 8, no. 3:117–22.

———. 1998. "The Preindustrial City as Ecosystem." *Capitalism, Socialism, Nature* 9, no. 1:105–10.

———. 1999. "The Classic Maya Collapse." *Capitalism, Socialism, Nature* 10, no. 1:81–90.

Hughes, Donald J., and J. V. Thirgood. 1982a. "Deforestation, Erosion, and Forest Management in Ancient Greece and Rome." *Journal of Forest History* 26, no. 2:60–75.

———. 1982b. "Deforestation in Ancient Greece and Rome: A Cause of Collapse." *Ecologist* 12:196–208.

Huntington, Ellsworth. 1907. *The Pulse of Asia*. London: Archibald Constable.

———. 1915. *Civilization and Climate*. New Haven: Yale University Press.

Huntington, Samuel. 1996. *The Clash of Civilizations and the Remaking of World Order*. New York: Simon and Schuster.

Hurst, Philip. 1990. *Rainforest Politics*. London: Zed Press.

Hutterer, Karl L., ed. 1977. "Economic Exchange and Social Interaction in Southeast Asia: Perspectives from Prehistory, History, and Ethnography." Michigan Paper on South and Southeast Asia, no. 13.

———. 1982. "Early Southeast Asia: Old Wine in New Skins?" *Journal of Asian Studies* 41:559–70.

Immerwahr, Sara A. 1960. "Mycenaean Trade and Colonization." *Archaeology* 13, no. 1:4–13.

Inglehart, Ronald. 1977. *The Silent Revolution: Changing Values and Political Styles among Western Publics*. Princeton: Princeton University Press.

———. 1981. "Post-Materialism in an Environment of Insecurity." *American Political Science Review* 75:880–900.

International Herald Tribune. 1999. August 11, 5.

International Task Force. 1985. *Tropical Forest Action Plan*. Washington D.C.: The World Bank.

Jacobsen, Jerome. 1986. "The Harappan Civilization: An Early State." In *Studies in the Archaeology of India and Pakistan*, edited by Jerome Jacobsen. New Delhi: Oxford University Press.

Jacobsen, Knut A. 1994. "The Institutionalization of the Ethics of 'Non-Injury' toward All 'Beings' in Ancient India." *Environmental Ethics* 16, no. 3:287–301.

Jacobsen, T. 1970. "On the Textile Industry at Ur under Ibbi-Sin." In *Toward the Image of Tammuz*, edited by W. L. Moran. Cambridge: Cambridge University Press.

Jacobsen, T., and R. M. Adams. 1958. "Salt and Silt in Ancient Mesopotamian Agriculture." *Science* 128:1251–8.

Jacques, C. 1979. "Funan, Zhenla: The Reality Concealed by the Chinese Views of Indochina." *Early S. E. Asia: Essays in Archaeology, History, and Historical Geography*, edited by R. B. Smith and W. Watson. New York: Oxford University Press.

James, P. E. 1932. "The Coffee Lands of Southeastern Brazil." *Geographical Review* 22:225–43.

James, P. J., I. J. Thorpe, N. Kokkiners. 1991. *Centuries of Darkness: A Challenge to the Conventional Chronology of Old World Archaeology*. London: Jonathan Cape.

Jarman, M. R., and H. N. Jarman. 1968. "The Fauna and Economy of Early Neolithic Knossos." *Annual of the British School in Athens* 63:741–64.

Johnson, D. L., and H. Gould. 1984. "The Effect of Climate Fluctuations on Human Populations: A Case Study of Mesopotamian Society." *Climate and Development*, edited by A. K. Biswas. Dublin: Tycooly International Publishing.

Jones, Alwyn. 1990. "Social Symbiosis: A Gaian Critique of Contemporary Social Theory." *The Ecologist* 20:108–13.

Jones, Glynis, Kenneth Wardle, Paul Halstead, Diane Wardle. 1986. "Crop Storage at Assiros." *Scientific American* 254, no. 3: 96–103.

Kaplan, Barbara. 1978. *Social Change in the Capitalist World-Economy*. Beverly Hills: Sage Publications.

Kardulias, Nick. 1996. "Multiple Levels in the Aegean Bronze Age World-System." *Journal of World-Systems Research* 2, no. 1:1–35.

Kardulias, Nick, and M. T. Shutes, eds. *Aegean Strategies: Studies of Culture and Environment on the European Fringe*. London: Rowman and Littlefield.

Kates, Robert, Jesse H. Ausubel, and Mimi Berbrian, eds. 1986. *Climate Impact Assessment: Studies of the Interactions of Climate and Society*. New York: John Wiley.

Kathirithamby-Wells, J. 1998. "Attitudes to Natural Resources and Environment among the Upland Forest and Swidden Community of Southeast Asia During the Nineteenth and Early Twentieth Centuries." In *Nature and the Orient*, edited by Richard Grove. New Delhi: Oxford University Press.

Kato, T. 1992. "Structural Changes in Japanese Forest Products Imported During the 1980s," *The Current State of Japanese Forestry*. Tokyo: Japanese Forest Economic Society.

Kelly, Petra. 1994. *Thinking Green: Essays on Environmentalism, Feminism, and Non-Violence*. Berkeley: Parallax Press.

Kempter, H., M. Gorres, and B. Frenzel. 1997. "Ti and Pb Concentrations in Rainwater-Fed Bogs in Europe as Indicators of Past Anthropogenic Activities." *Water, Air and Soil Pollution* 100:367–77.

Kennedy, J. 1898. "Early Commerce of Babylon with India." *Journal of the Asiatic Society of Bengal*, 241–88.

Kennedy, K. A. R., and G. Possehl, eds. 1976. "Ecological Backgrounds of South Asian Prehistory." *Occasional Papers and Theses of the South Asian Program of Cornell University.* Ithaca: Cornell University.

Kenoyer, Jonathan. 1989. "Socio-Economic Structures of the Indus Civilization as Reflected in Special Crafts and the Question of Ritual Segregation." In *Old Problems and New Perspectives in the Archaeology of South Asia*, edited by J. M. Kenoyer. Vol. 2 of Wisconsin Archaeological Reports.

———. 1993. "Excavations on Mound E, Harappa: A Systematic Approach to the Study of Indus Urbanization." In *South Asian Archaeology*, edited by A. J. Gail and G. J. R. Morisson. Stuttgart: F. S. Verlag.

———. 1994. "Carnelian Bead Production in Khmabat, India." In *Living Traditions*, edited by B. Allchin. New Delhi: IBH Publishing.

———. 1997a. "Trade and Technology of the Indus Valley: New Insights from Harappa, Pakistan." *World Archaeology* 29, no. 2:262–80.

———. 1997b. "Early City-States in South Asia: Comparing the Harappan Phase and the Early Historic Period." In *The Archaeology of City-States: Cross Cultural Approaches*, edited by D. L. Nichols and T. H. Charlton. Washington, D.C.: Smithsonian Institution Press.

Kerr, Richard. 1998. "Sea Floor Dust Shows Drought Felled Akkadian Empire." *Science* 279:325–26.

Kheel, Marti. 1990. "Ecofeminism and Deep Ecology: Reflections on Identity and Difference." In *Reweaving the World: The Emergence of Ecofeminism*, edited by I. Diamond. San Francisco: Sierra Club.

Kick, E., Thomas J. Burns, Byron Davis, David A. Murray, and Dixie A. Murray. 1996. "Impacts of Domestic Population Dynamics and Foreign Wood Trade on Deforestation: A World-Systems Perspective." *Journal of Developing Societies* 12:68–87.

Kirch, Patrick V. 1997. "Microcosmic Histories." *American Anthropology* 99, no. 1:30–42.

Knapp, A. Bernard. 1993. "Thalassocracies in the Bronze Age Eastern Mediterranean Trade: Making and Breaking a Myth." *World Archaeology* 24, no. 3:332–47.

Kohl, Philip. 1978. "The Balance of Trade in Southwestern Asia in the Mid-Third Millennium." *Current Anthropology* 193:480–81.

———. 1987a. "The Ancient Economy, Transferable Technologies, and the Bronze Age World System: A View from the Northeastern Frontiers of the Ancient Near East." In *Centre and Periphery in the Ancient World*, edited by Michael Rowlands, Mogens Larsen, Kristian Kristiansen. New York: Cambridge University Press.

———. 1987b. "Sumer and Indus Valley Civilization Compared: Towards an Historical Understanding of the Evolution of Early States." In *Perspectives in U.S. Marxist Anthropology*, edited by D. Hakken and H. Lessinger. London: Westview Press.

Kolb, Albert. 1971. *East Asia*. London: Methuen.

Krabb, Wolfgang R. 1974. *Gesellschaftverauderung durch Lebensreform*. Gottingen: Vandenhoeck and Ruprecht.

Kraft, John, Stanley E. Aschenbrenner, and George Rapp, Jr. 1977. "Paleographic Reconstructions of Coastal Aegean Archaeological Sites." *Science* 195:941–43.

Kramer, Samuel Noah. 1981. *History Begins at Sumer*. Philadelphia: University of Pennsylvania Press.

Kristiansen, Kristian. 1993. "The Emergence of the European World System in the Bronze Age: Divergence, Convergence, and Social Evolution During the First and Second Millennia B.C. in Europe." Sheffield Archaeological Monographs, no. 6.

———. 1998. *Europe before History*. Cambridge: Cambridge University Press.

Kuniholm, P. I. 1992. "Dendrochronological Wood from Anatolia and Environs." In *Trees and Timber in Mesopotamia*, edited by N. Postgate and T. Powell. Bulletin on Sumerian Agriculture 6. Cambridge: Cambridge University.

Kuniholm, P. I., and C. L. Striker. 1987. "Dendrochronological Investigations in the Aegean and Neighboring Regions 1983–1986." *Journal of Field Archaeology* 14:385–98.

Ladurie, Emmanuel Le Roy. 1971. *Times of Feast and Times of Famine*. New York: Doubleday.

Lal, B. B. 1993. "A Glimpse of the Social Stratification and Political Set-Up of the Indus Civilization." In *Harappan Studies*, vol. 1, edited by G. Possehl and M. Tosi. New Delhi: Oxford University Press.

———. 1997. *The Earliest Civilization of South Asia: Rise, Maturity, and Decline*. New Delhi: Aryan Books International.

Lamb, H. 1967. "R. Carpenter's Discontinuity in Greek Civilization." *Antiquity* 41:233–34.

———. 1968. *The Changing Climate*. London: Methuen.

———. 1982a. *Climate, History, and the Modern World*. London: Methuen.

———. 1982b. "Reconstruction of the Course of Postglacial Climate over the World." In *Climate Change in Later Pre-History*, edited by Anthony Harding. Edinburgh: Edinburgh University Press.

Lamberg-Karlovsky, C. C. 1975. "Third Millennium Exchange and Production." In *Ancient Civilization and Trade*, edited by Jeremy Sabloff and C. C. Lamberg-Karlovsky. Albuquerque: University of New Mexico Press.

———. 1979. "Trade Mechanisms in Indus-Mesopotamian Interrelations." In *Ancient Cities of the Indus*, edited by G. Possehl. New Delhi: Vikas.

———. 1986. "Third-Millennium Structure and Process: From the Euphrates to the Indus and the Oxus to the Indian Ocean." *Oriens Antiquvvs* 25:189–219.

Lambrick, H. T. 1967. "The Indus Flood Plain and the 'Indus' Civilization." *The Geographical Journal* 133, no. 4:483–89.

Lane, Frederic. 1934. *Venetian Ships and Shipbuilders of the Renaissance*. Baltimore: Johns Hopkins University Press.

Lange, J. L. 1990. "Refusal to Compromise: The Case of Earth First!" *Western Journal of Speech Communication* 54:473–94.

Lansky, Mitch. 1992. *Beyond the Beauty Strip*. Boston: Tilbury House.

LaPiere, R. T. 1965. *Social Change*. New York: McGraw Hill.

Laurence, Ray. 1994. *Roman Pompeii: Space and Society*. London: Routledge.

Leahy, Joe. 1999. "Indonesia Depends on Illegal Logging." *(London) Financial Times*, December 17, 5.

Lee, J. A., and J. H. Talle. 1973. "Regional and Historical Aspects of Lead Pollution in Britain." *Nature* 245:216–18.

Leemans, W. F. 1960. *Foreign Trade in the Old Babylonian Period as Revealed by Texts from Southern Mesopotamia*. Leiden: Brill.

Legge, A. J. 1972. "Prehistoric Exploitation of the Gazelle in Palestine." In *Papers in Economic Prehistory*, edited by E. S. Higgs. Cambridge: Cambridge University Press.

Leiss, William. 1974. *The Domination of Nature*. Boston: Beacon Press.

Leong, S. H. 1990. "Collecting Centres, Feeder Points and Entrepôts in the Malay Peninsula 1000 B.C.–A.D. 1400." In *The Southeast Asian Port and Polity*, edited by J. Kathirithamby-Wells and J. Villiers. Singapore: Singapore University Press.

Leopold, Aldo. 1949. *A Sand County Almanac*. New York: Oxford University Press.

Leshnik, L. S. 1973. "Land Use and Ecological Factors in Prehistoric North-West India." In *South Asian Archaeology*, edited by Norman Hammond. Park Ridge, N.J.: Noyes Press.

Lian, Francis. 1988. "The Economies and Ecology of the Production of the Tropical Rainforest Resources by Tribal Groups of Sarawak, Borneo." In *Changing Tropical Forests*, edited by J. Dargavel. Canberra, Aus.: Centre for Resource and Environmental Studies.

Lim, H. F. 1992. "Aboriginal Communities and the International Trade in Non-Timber Forest Products: The Case of Peninsular Malaysia." in *Changing Pacific Forests*, edited by J. Dargavel and R. Tucker. Durham: Forest History Society.

Linse, Ulrich. 1983. *Barfuessuge Propheten:Erloeser der 20er Jahse*. Berlin: Siedler Verlag.

Lipietz, Alain. 1992. *Towards a New Economic Order*. Oxford: Oxford University Press.

———. 1995. *Green Hopes*. London: Polity Press.

List, Peter, ed. 1993. *Radical Environmentalism*. Belmont, Cal.: Wadsworth.

Lo, Tung-Pang. 1958. "The Decline of the Early Ming Navy." *Oriens Extremus* 5:149–68.

Long, N. 1990. "From Paradigm Lost to Paradigm Regained? The Case for an Action-Oriented Sociology of Development." *European Review of Latin American and Caribbean Studies* 49:3–24.

Lovelock, J. E. 1979. *Gaia: A New Look at Life on Earth*. Oxford: Oxford University Press.

Lowdermilk, W. G. 1925. "Relation of Deforestation, Slope Cultivation, Erosion, and Overgrazing to Increased Silting of the Yellow River." *American Forests and Forest Life* 379:25–45.

Lower, Arthur. 1973. *Great Britain's Woodyard*. Kingston: Queen's University Press.

———. 1977. *Colony to Nation*. Toronto: McClelland and Stewart.

Lyotard, J. 1984. *The Postmodern Condition*. Minneapolis: University of Minnesota Press.

Mabbett, I. W. 1977. "The Indianization of S.E. Asia: Reflections on the Historical Sources." *Journal of Southeast Asian Studies* 8, no. 2:143–61.

Mackay, E. J. H. 1931. "Further Links between Ancient Sind, Sumer, and Elsewhere." *Antiquity* 5:459–73.

MacNeill, Jim, Pieter Winsemius, and Taizo Yakushiji. 1991. *Beyond Interdependence: The Meshing of the World's Economy and the Earth's Ecology*. New York: Oxford University Press.

Mahlman, J. 1997. "Uncertainties in Projections of Human-Caused Climate Warming." *Science* 278:1416–18.

Malik, M. 1968. *Indian Civilization: The Formative Period*. Simla, India: IIAS.

Manes, Chris. 1990. *Green Rage*. Boston: Little Brown.

Manguin, Pierre-Yves. 1985. "Research on the Ships of Srivijaya." In *SEAMEO Project in Archaeology and Fine Arts*. Consultative Workshop on Archaeological and Environmental Studies on Srivijaya. September 16–30.

Mann, Michael. 1986. *The Sources of Social Power*. Vol. 1. Cambridge: Cambridge University Press.

———. 1993. *The Rise of Classes and Nation-States*. Vol. 2. Cambridge: Cambridge University Press.

———. 1998. "Ecological Change in North India." In *Nature and the Orient*, edited by Richard Grove. New Delhi: Oxford University Press.

Manning, Stuart W. 1994. "The Emergence of Divergence: Development and Decline in Bronze Age Crete and the Cyclades." In *Development and Decline in the Mediterranean Bronze Age*, edited by C. Mathers and S. Stoddatm. Sheffield, Eng.: J. R. Collins.

Marchak, Pat. 1995. *Logging the Globe*. Kingston: Queen's University Press.

Marcuse, Herbert. 1956. *Eros and Civilization*. Boston: Beacon Press.

———. 1992. "Ecology and the Critique of Modern Society." *Capitalism, Nature, Socialism* 3, no. 3:29–38.

Marinatos, S. 1939. "The Volcanic Eruption of Minoan Crete." *Antiquity* 13:425–39.

Marks, Robert B. 1996. "Commercialization without Capitalism: Processes of Environmental Change in South China 1550–1850." *Environmental History* 1, no. 1:56–82.

———. 1998. *Tigers, Rice, Silk and Silt: Environment and Economy in Late Imperial South China*. Cambridge: Cambridge University Press.

Marr, David G., and A. C. Milner, eds. 1990. *Southeast Asia in the Ninth to Fourteenth Centuries*. Singapore: Institute of Southeast Asian Studies.

Marsh, George. 1895. *Man and Nature*. New York: Charles Scribner.

Marshall, J. 1931. *Mohenjo-Daro and the Indus Civilization*. London: Cambridge University Press.

Martin, William. 1994. "The World-Systems Perspective in Perspective: Assessing the Attempt to Move Beyond Nineteenth-Century Eurocentric Conceptions." *Review* 17, no. 2:145–86.

Maser, Chris. 1992. *Global Imperative*. Walpole, N.H.: Stillpoint.

Massa, Ilmo. 1999. "The Development of the Risk Economy in the Circumpolar North." In *Ecology and the World-System*, edited by Walter Goldfran, David Goodman, and Andrew Szasz. Westport, Conn.: Greenwood Press.

Mathers, C., and S. Stoddart, eds. *Development and Decline in the Mediterranean Bronze Age*. Sheffield, Eng.: John Collins Publications.

Mathews, F. 1988. "Conservation and Self-Realization: A Deep Ecology Perspective." *Environmental Ethics* 10:34–46.

McC Adams, R. 1966. *The Evolution of Urban Society*. Chicago: Aldine.

———. 1981. *Heartland of Cities*. Chicago: University of Chicago Press.

———. 1996. *Paths of Fire*. Princeton: Princeton University Press.

McCoy, F. 1996. "Climatic Change in the Eastern Mediterranean Area During the Past 240,000 Years." In *Thera and the Aegean World II*, edited by C. Doumas. Athens: G. Tsiveriotis.

McKibben, Bill. 1989. *The End of Nature*. New York: Random House.

McLaughlin, Andrew. 1993. *Regarding Nature*. Albany: State University of New York Press.

McMichael, Philip. 1984. *Settlers and the Agrarian Question*. Cambridge: Cambridge University Press.

McNeill, J. R. 1998. "China's Environmental History." In *Sediments of Time*, edited by Mark Elvin. Cambridge: Cambridge University Press.

McNeill, William. 1962. *The Rise of the West*. Chicago: University of Chicago Press.

———. 1974. *Venice: The Hinge of Europe 1081–1797*. Chicago: University of Chicago Press.

———. 1976. *Plagues and Peoples*. New York: Anchor Books.

———. 1990. "The Rise of the West after Twenty-Five Years." *Journal of World History* 1:1–21.

———. 1992. *The Global Condition*. Princeton: Princeton University Press.

Meadows, Donella, Dennis L. Meadows, and Jorgen Randers. 1992. *Beyond the Limits: Confronting Global Collapse, Envisioning a Sustainable Future*. Post Mills, Vt.: Chelsea Green.

Meiggs, R. 1982. *Trees and Timber of the Ancient Mediterranean World*. Oxford: Clarendon Press.

Melucci, Alberto. 1985. "The Symbolic Challenge of Contemporary Movements." *Social Research* 52, no. 4:789–816.

———. 1989. *Nomads of the Present*. Philadelphia: Temple University Press.

Mendez, Chico. 1990. *Fight for the Forest*. London: Latin American Bureau.

Menzies, Nicholas. 1992. "Sources of Demand and Cycles of Logging in Premodern China." In *Changing Pacific Forest*, edited by John Dargavel and Richard Tucker. Durham: Forest History Society.

———. 1996. "Forestry." In *Science and Civilization in China*, vol. 6, edited by Joseph Needham. Cambridge: Cambridge University Press.

Merchant, Carolyn. 1990. "Ecofeminism and Feminist Theory." In *Reweaving the World*, edited by I. Diamond and G. Orenstein. San Francisco: Sierra Club.

———. 1992. *Radical Ecology*. New York: Routledge.

Mies, Maria, and V. Shiva. 1993. *Ecofeminism*. London: Zed Press.

Miller, G. 1988. *Living in the Environment*. Belmont, Cal.: Wadsworth.

Miller, Innes. 1969. *The Spice Trade of the Roman Empire*. Oxford: Clarendon Press.

Mitchell, John G. 1991. *Dispatches from the Deep Woods*. Lincoln: University of Nebraska Press.

Modelski, George, and William Thompson. 1996. *Global Sectors and World Powers*. Columbia, S.C.: University of South Carolina Press.

Moeller, W. O. 1976. *The Wool Trade of Ancient Pompeii*. London: Brill.

Moody, J. A. 1997. "The Cretan Environment: Abused or Just Misunderstood?" In *Aegean Strategies: Studies of Culture and Environment on the European Fringe*, edited by P. N. Kardulias. Lanham, Md.: Rowman and Littlefield.

Moorey, P. R. S. 1994. *Ancient Mesopotamian Materials and Industries*. Oxford: Clarendon Press.

Moran, E. F. 1982. "Ecological, Anthropological, and Agronomic Research in the Amazon Basin." *Latin American Studies* 17:3–41.

Mori No Koe. 1993. *Japan and the World's Forests* newsletter. Tokyo.

Morner, N. A., and W. Karlen, eds. 1984. *Climate Changes on a Yearly to Millennial Basis*. Boston: D. Reidel.

Morris, Ian. 1987. *Burial and Ancient Society: The Rise of the Greek City-State*. Cambridge: Cambridge University Press.

———. 1997. "An Archaeology of Equalities? The Greek City-States." In *The Archaeology of City-States: Cross-Cultural Approaches*, edited by D. L. Nichols and T. H. Charlton. Washington, D.C.: Smithsonian Institution Press.

Muhly, J. D. 1973. "Tin Trade Routes of the Bronze Age." *American Scientist* 61, no. 4:404–13.

Muir, John. 1911. *The Mountains of California*. New York: The Century Co.

———. 1916. *A Thousand Mile Walk to the Gulf*. Boston, Mass.: Houghton, Mifflin Co.

——. 1986. *The Yosemite*. Madison: University of Wisconsin Press.

Murphey, Rhoads. 1951. "The Decline of North Africa since the Roman Occupation: Climatic or Human?" *Annals of the Association of American Geographer* 41, no. 1:116–32.

——. 1983. "Deforestation in Modern China." In *Global Deforestation and the Nineteenth-Century World Economy*, edited by Richard Tucker and John F. Richards. Durham, N.C.: Duke Press Policy Studies.

Mylonas, George. 1966. *Mycenae and the Mycenaean Age*. Princeton: Princeton University Press.

Naess, Arne. 1973. "The Shallow and the Deep, Long-Range Ecology Movements: A Summary." *Inquiry* 16:95–100.

——. 1984. "A Defense of the Deep Ecology Movement." *Environmental Ethics* 6:265–70.

——. 1985. "Identification as a Source of Deep Ecological Attitudes." In *Deep Ecology*, edited by M. Tobais. San Diego: Avant Books.

——. 1986. "The Deep Ecological Movement: Some Philosophical Aspects." *Philosophical Inquiry* 8:10–31.

——. 1987. "Self-Realization: An Ecological Approach to Being in the World." *The Trumpeter* 4, no 3:17–34.

——. 1989. *Ecology, Community, and Lifestyle*. Cambridge: Cambridge University Press.

——. 1995. "Politics and the Ecological Crisis." In *Deep Ecology for the Twenty-First Century*, edited by George Sessions. Boston: Shambala.

Nash, Daphne. 1987. "Imperial Expansion under the Roman Republic." In *Centre and Periphery in the Ancient World*, edited by Michael Rowlands, Mogens Larsen, and Kristian Kristiansen. New York: Cambridge University Press.

Nash, R. F. 1982. *The Rights of Nature*. Madison: University of Wisconsin Press.

National Academy of Sciences. 1980. *Lead in the Human Environment*. Washington, D.C.: U.S. National Academy of Sciences Commission on Natural Resources.

Nectoux, Francois, and Yoichi Kuroda. 1990. *Timber from the South Seas*. Geneva: World Wildlife Fund.

Neumann, J., and S. Parpola. 1987. "Climatic Change and the Eleventh–Tenth Century Eclipse of Assyria and Babylonia." *Journal of Near Eastern Studies* 46:161–82.

Neumann, J., and R. Sigrist. 1978. "Harvest Dates in Ancient Mesopotamia as Possible Indicators of Climatic Variations." *Climate Change* 1:239–52.

Niehues-Probsting, Heinrich. 1979. *Der Uynismus des Diogenes und der Begriff des Zynismus*. Munchen: Wilhelm Fink Verlag.

Nilsson, Sten, and Anatoly Shivdenko. 1997. "The Russian Forest Sector." Background paper for World Commission on Forests and Sustainable Development. International Institute for Applied Systems Analysis.

Noble, Ian, and Rodolfo Dirzo. 1997. "Forests as Human-Dominated Ecosystems." *Science* 277:522–25.

Norberg-Hodge, Helena. 1991. *Ancient Futures*. San Francisco: Sierra Club.

Nriagu, J. O. 1983. *Lead and Lead Poisoning*. New York: Wiley.

Oates, D. 1968. *Studies in the Ancient History of Northern Iraq*. Oxford: Oxford University Press.

Oates, D., and J. Oates. 1976. *The Rise of Civilization*. Oxford: Elsevier.

O'Briant, Walter. 1974. "Man, Nature and the History of Philosophy." In *Philosophy and Environment Crisis*, edited by W. Blackstone. Athens: University of Georgia Press.

O'Connor, D. 1983. "New Kingdom and Third Intermediate Period, 1552–664 B.C." In *Ancient Egypt: A Social History*, edited by B. G. Trigger, B. J. Kemp, D. O'Connor, and A. B. Lloyd. Cambridge: Cambridge University Press.

O'Connor, James. 1988. "Capitalism, Nature, Socialism: A Theoretical Introduction." *Capitalism, Nature, Socialism* 1:11–38.

———. 1991. "The Second Contradiction of Capitalism." *Capitalism, Nature, Socialism*, pamphlet no. 1.

Oldenstein, J. 1985. "Manufacture and Supply of the Roman Army with Bronze Fittings." In *The Production and Distribution of Roman Military Equipment*, edited by M. C. Bishop. Oxford: British Archaeological Reports International Series, no. 275.

Oppenheim, A. L. 1964. *Ancient Mesopotamia*. Chicago: University of Chicago Press.

———. 1979. "The Seafaring Merchants of Ur." In *Ancient Cities of the Indus*, edited by G. Possehl. New Delhi: Vikas.

O'Riordan, Timothy. 1981. *Environmentalism*. London: Pion.

Osborne, Anne. 1998. "Highlands and Lowlands." In *Sediments of Time*, edited by Mark Elvin. Cambridge: Cambridge University Press

Packer, James E. 1988. "Roman Building Techniques." In *Civilization of the Ancient Mediterranean*, edited by M. Grant and Grant Kitzinger. New York: Charles Scribner and Sons.

Parker, Thomas S. 1984. "Exploring the Roman Frontier in Jordan." *Archaeology* 37, no. 5:33–39.

Peluso, Nancy Lee. 1992a. "A Short History of State Forest Management in Java." In *Keepers of the Forest*, edited by Mark Poffenberger. West Hartford, Conn.: Kumarian.

———. 1992b. *Rich Forests, Poor People: Resources Control and Resistance in Java*. Berkeley: University of California Press.

Penna, Ian. 1992. *Japan's Paper Industry*. Tokyo: Chikyu no Tomo.

Pepper, D. 1984. *The Roots of Modern Environmentalism*. London: Croom Helm.

Perdue, Peter. 1987. *Exhausting the Earth: State and Peasant in Hunan 1500–1850*. Cambridge: Cambridge University Press.

Perlin, John. 1989. *A Forest Journey*. Cambridge: Harvard University Press.

Petesch, Patti. 1990. *Tropical Forests: Conservation with Development*. Washington, D.C.: Overseas Development Council.

Pezzey, John. 1992. "Sustainable Development Concept: An Economic Analysis." World Bank Environment Paper no. 2. Washington D.C.

Pickett, Karen. 1993. "Redwood Summer Retrospective." In *Radical Environmentalism*, edited by Peter List. Belmont, Cal.: Wadsworth.

Piggott, Stuart. 1950. *Pre-Historic India*. Harmondsworth, Eng.: Penguin.

Pires, Tome. 1944. *The Suma Oriental of Tome Pires*. London: Hakluyt Society.

Plumwood, Val. 1992. "Feminism and Ecofeminism: Beyond the Dualistic Assumptions of Women, Men, and Nature." *The Ecologist* 22, no. 1:8–13.

Poczy, Klara. 1980. "Pannonian Cities." In *The Archaeology of Roman Pannonia*, edited by A. Lengyel and E. T. B. Rudan. Lexington: University Press of Kentucky.

Polanyi, Karl. 1957. *The Great Transformation*. Boston: Beacon Press.

Pomeranz, Kenneth. 1994. *The Making of a Hinterland*. Berkeley: University of California Press.

Ponting, Clive. 1991. *A Green History of the World*. London: Penguin.

Porritt, Jonathan. 1990. Foreword to *Tropical Rainforest Politics*, by P. Hurst. London: Zed Press.

Possehl, G., ed. 1979. *Ancient Cities of the Indus*. New Delhi: Vikas.

——. 1982. "The Harappan Civilization: A Contemporary Perspective." In *Harappan Civilization: A Contemporary Perspective*, edited by G. Possehl. New Delhi: Oxford University Press.

——, ed. 1993. *Harappan Civilization: A Recent Perspective*. New Delhi: Oxford University Press.

——. 1996. "Meluhha." In *The Indian Ocean in Antiquity*, edited by Julian Reade. London: Kegan and Paul.

——. 1997. "Climate and the Eclipse of the Ancient Cities of the Indus." In *Third Millennium B.C. Climate Change and Old World Collapse*, edited by H. Nuzhet Dalfes, George Kukla, and Harvey Weiss. Heidelberg: Springer-Verlag.

Possehl, G., and M. Raval. 1989. *Harappan Civilization and the Rojdi*. New Delhi: Oxford.

Possehl, G., and M. Tosi, eds. 1993. *Harappan Studies*. New Delhi: Oxford University Press.

Postgate, J. N. 1992. *Early Mesopotamia: Society and Economy at the Dawn of History*. London: Croom Helm.

Postgate, J. N., and M. A. Powell. 1992. "Trees and Timber in Assyrian Texts." In *Trees and Timber in Mesopotamia*, edited by J. N. Postgate and M. A. Powell. Bulletin on Sumerian Agriculture 6. Cambridge: Cambridge University.

Potter, Lesley. 1988. "Indigenes and Colonizers: Dutch Forest Policy in South and East Brunei (Kalimantan) 1900 to 1950." In *Changing Tropical Forests*, edited by John Dargavel. Canberra: Australian National University.

Potts, D. T. 1977. *Mesopotamian Civilization: The Material Foundations*. Ithaca, N.Y.: Cornell University Press.

Powell, M. A. 1992. "Timber Production in PreSargonic Lagash." In *Trees and Timber in Mesopotamia*, edited by J. N. Postgate and M. A. Powell. Bulletin on Sumerian Agriculture 6. Cambridge: Cambridge University.

Proctor, James. 1995. "Whose Nature? Contested Moral Terrain of Ancient Forests." In *Uncommon Ground*, edited by William Cronon. New York: Norton.

Rahman, Nik. 1990. "The Kingdom of Srivijaya as Social, Political, and Cultural Entity." In *The Southeast Asian Port and Polity*, edited by J. Kathirithamby-Wells and John Villiers. Singapore: National University of Singapore.

Raikes, Robert. 1964. "The End of the Ancient Cities of the Indus." *American Anthropologist* 66, no. 2:284–99.

Raikes, Robert, and G. F. Dales. 1977. "The Mohenjo-Daro Floods Reconsidered." *Journal of the Palaeontological Society of India* 20:251–60.

Raikes, Robert, and R. Dyson. 1961. "The Prehistoric Climate of Baluchistan and the Indus Valley." *American Anthropologist* 63, no. 2:265–81.

Randsborg, Klavs. 1991. *The First Millennium A.D. in Europe and the Mediterranean*. Cambridge: Cambridge University Press.

Ratnagar, Shereen. 1981. *Encounters: The Westerly Trade of the Harappan Civilization*. New Delhi: Oxford University Press.

——. 1986. "An Aspect of Harappan Agricultural Production." *Studies in History* 2:137–53.

——. 1991. *Enquiries into the Political Organization of the Harappan Society*. Pune: Ravish.

———. 1994. "Harappan Trade in its 'World' Context." *Man and Environment* 19, nos. 1–2:115–27.

Ray, Himanshu Prabha. 1989. "Early Maritime Contacts between South and Southeast Asia." *Journal of Southeast Asian Studies* 20, no. 1:42–54.

Redman, Charles. 1978. *The Rise of Civilization*. San Francisco: W. H. Freeman.

Reed, Peter, and D. Rothenburg, eds. 1992. *Wisdom in the Open Air*. Minneapolis: University of Minnesota Press.

Reid, A. 1992. "Economic and Social Change." In *Cambridge History of Southeast Asia*, edited by N. Tarling. Cambridge: Cambridge University Press.

———. 1998. "Humans and Forests in Pre-Colonial Southeast Asia." In *Nature and the Orient*, edited by Richard Grove. Bombay: Oxford University Press.

Renberg, I. 1994. "Preindustrial Atmospheric Lead Contamination Detected in Swedish Lake Sediments." *Nature* 368: 323–26.

Renfrew, Colin. 1972. *The Emergence of Civilization*. London: Collins.

———. 1979. "Systems Collapse as Social Transformation: Catastrophe and Anastrophe in Early State Society." In *Transformations: Mathematical Approaches to Cultural Change*, edited by C. Renfrew and K. L. Cooke. New York: Academic Press.

Research Working Group. 1979. "Cyclical Rhythms and Secular Trends of the Capitalist World-Economy: Some Premises, Hypotheses and Questions." *Review* 2, no. 4:483–500.

Rice, Michael. 1994. *Archaeology of the Arabian Gulf*. London: Routledge.

Richards, J. F., and M. McAlpin. 1983. "Cotton Cultivating and Land Clearing in the Bombay Deccan and Karnatak 1818–1920." In *Global Deforestation and the Nineteenth-Century World Economy*, edited by Richard Tucker and J. F. Richards. Durham, N.C.: Duke Press Policy Studies.

Roberts, N. 1989. *The Holocene: An Environmental History*. Oxford, Eng.: Basil Blackwell.

Rockham, O. J., and J. A. Moody. 1996. *The Making of the Cretan Landscape*. Manchester, Eng.: Manchester University Press.

Rosch, Manfred. 1990. "Vegetationsgechichtliche Untersuchungen im Durchenbergried." In *Siedlungsarchaologie im Alpenvorland II*, edited by Andre Billamboz. Stuttgart: Konrad Theiss.

———. 1996. "New Approaches to Prehistoric Land-Use Reconstruction in Southwestern Germany." *Vegetation History and Archaeobotany* 5, no. 1/2:45–57.

———. 1998. "The History of Crops and Crop Weeds in Southwestern Germany from the Neolithic Period to Modern Times as Shown by Archaeological Evidence." *Vegetation History and Archaeobotany* 7, no. 2:65–78.

Roth, Dennis. 1983. "Philippine Forests and Forestry 1565–1920." In *Global Deforestation and the Nineteenth-Century World Economy*, edited by Richard P. Tucker and John Richards. Durham, N.C.: Duke Press Policy Studies.

Rothenburg, David. 1992. *Is It Painful to Think?* Minneapolis: University of Minnesota Press.

Rowlands, Michael, Mogens Larsen, and Kristian Kristiansen, eds. 1987. *Centre and Periphery in the Ancient World*. New York: Cambridge University Press.

Rowton, M. B. 1967. "The Woodlands of Ancient Western Asia." *Journal of Near Eastern Studies* 26:261–77.

Runnels, Curtis N. 1995. "Environmental Degradation in Ancient Greece." *Scientific American* 272, no. 3:96–99.

Rush, James. 1991. *The Last Tree: Reclaiming the Environment in Tropical Asia*. New York: Asia Society.

Russell, J. C. 1958. "Late Ancient and Medieval Population." *Transactions of the Ancient Philosophical Society* 48, part 3.

Russell, James. 1980. "Anemurium: The Changing Face of a Roman City." *Archaeology* 33, no. 5:31–40.

Russell, William. 1969. *Man, Nature, and History*. Garden City, N.J.: Natural History Press.

Ryerson, Stanley B. 1975. *The Founding of Canada*. Toronto: Progress Books.

Saggs, H. W. F. 1962. *The Greatness That Was Babylon*. New York: Hawthorn Books.

——. 1989. *Civilization before Greece and Rome*. New Haven: Yale University Press.

Sahni, M. R. 1956. "Biogeological Evidence Bearing on the Decline of the Indus Valley Civilization." *Journal of the Palaeontological Society of India* 1, no. 1:101–7.

Sallares, Robert. 1991. *The Ecology of the Ancient Greek World*. Ithaca: Cornell University Press.

Salleh, Ariel. 1964. "Deeper than Deep Ecology: The Ecofeminist Connection." *Environmental Ethics* 6: 339–45.

Salt, Henry Stephens. 1980. "Animals' Rights: Considered in Relation to Social Progress." Clarks Summit Society for Animal Rights Inc.

Sandhu, K., and P. Wheatley. 1989. *Melaka: The Transformation of a Malay Capital 1400–1980*. New York: Oxford University Press.

Sandweiss, D. 1996. "Geoarchaeological Evidence from Peru for a Five Thousand–Year B.P. Onset of El Niño." *Science* 273:1531–33.

Scarre, C., and B. Fagan. 1997. *Ancient Civilizations*. London: Longman.

Schaeffer, C. F. A. 1948. *Stratigraphie comparee et chronologie de l'Asie occidentale (III et II millenaire)*. Oxford: Oxford University Press.

Schafer, Edward H. 1962. "Conservation of Nature under the Tang Dynasty." *Journal of Economics and Social History of the Orient* 5:279–308.

——. 1963. *The Golden Peaches of Samarkand*. Berkeley: University of California Press.

——. 1967. *The Vermilion Bird*. Berkeley: University of California Press.

Schmidt, Alfred. 1971. *The Concept of Nature in Marx*. London: New Left Books.

Schnaiberg, Allan. 1980. *The Environment from Surplus to Scarcity*. New York: Oxford University Press.

Schnaiberg, Allan, and K. Gould. 1994. *Environment and Scarcity: The Enduring Conflict*. New York: St. Martin's Press.

Schneider, J. 1977. "Was There a Pre-Capitalist World-System?" *Peasant Studies* 6, 1:20–29.

Schofield, Elizabeth. 1982. "The Western Cyclades and Crete: A 'Special' Relationship." *Oxford Journal of Archaeology* 1:9–25.

——. 1984. "Coming to Terms with Minoan Colonists." In *The Minoan Thalassocracy*, edited by R. Hagg and N. Marinatos. Stockholm: Paul Astroms Forlag.

Semple, Ellen Churchill. 1931. *The Geography of the Mediterranean Region: Its Relation to Ancient History*. New York: Henry Holt.

Sessions, George. 1985. "Western Process Metaphysics." In *Deep Ecology*, edited by Bill Devall and George Sessions. Salt Lake City: Peregrine Books.

——. 1987. "The Deep Ecology Movement: A Review." *Environmental Review* 11:105–12.

Shaffer, J. 1982. "Harappan Culture: A Reconsideration." In *Harappan Civilization: A Contemporary Perspective*, edited by G. Possehl. New Delhi: Oxford University Press.

Shaw, B. 1981. "Climate, Environment, and History: The Case of Roman North Africa." In *Climate and History*, edited by T. M. L. Wigley, M. J. Ingram, and G. Farmer. Cambridge: Cambridge University Press.

———. 1995. *Environment and Society in Roman North Africa*. Brookfield, Mass.: Variorum.

Shaw, Norman. 1914. *Chinese Forest Trees and Timber Supply*. London: Unwin.

Shendge, M. 1977. *The Civilized Demons: The Harappans in Rgveda*. New Delhi: Abhinav Publications.

Shennan, Stephen. 1993. "Settlement and Social Change in Central Europe 3500–1500 B.C." *Journal of World Prehistory* 7, no. 2:121–61.

Sheratt, Andrew. 1992. "What Would a Bronze Age World System Look Like? Relations between Temperate Europe and the Mediterranean in Later Prehistory." *Journal of European Archaeology* 1, no. 2:1–57.

Sheratt, Andrew, and Susan Sheratt. 1991. "From Luxuries to Commodities: The Nature of the Mediterranean Bronze Age Trading Systems." In *Bronze Age Trade in the Mediterranean*, edited by N. Gale. Jonsered: Paul Astroms Forlag.

Sheratt, Susan, and Andrew Sheratt. 1993. "The Growth of the Mediterranean Economy in the Early First Millennium." *World Archaeology* 24, no. 3:361–78.

Sidebotham, Steven. 1991. "Parts of the Red Sea and the Arabia-India Trade." In *Rome and India: The Ancient Sea Trade*, edited by V. Regley and R. DePuma. Madison: University of Wisconsin Press.

Silver, Morris. 1985. *Economic Structures of the Ancient Near East*. London: Croom Helm.

Slater, David. 1990. "Fading Paradigm and New Agendas: Crisis and Controversy in Development Studies." *European Review of Latin American and Caribbean Studies* 49:25–32.

Smil, Vaclav. 1984. *The Bad Earth: Environmental Degradation in China*. New York: M.E. Sharpe.

———. 1993. *China's Environmental Crisis: An Inquiry into the Limits of National Development*. Armonk, N.Y.: M.E. Sharpe.

Smith, Brian K. 1989. *Reflections on Resemblance, Ritual, and Religion*. New York: Oxford University Press.

———. 1990. "Eaters, Food and Social Hierarchy in Ancient India: A Dietary Guide to a Revolution of Values." *Journal of the American Academy of Religion* 58, no. 2:177–205.

Smith, Stuart Tyson. 1995. *Askut in Nubia: The Economics and Ideology of Egyptian Imperialism in the Second Millennium B.C.* London: Kegan Paul International.

Snodgrass, A. M. 1971. *The Dark Age of Greece*. Edinburgh: University Press.

———. 1989. "The Coming of the Iron Age in Greece: Europe's Earliest Bronze/Iron Transition." In *The Bronze-Iron Age Transition in Europe*, edited by M. L. S. Sorensen and R. Thomas. Oxford: BAR Int. Sr. 483.

Snyder, Gary. 1974. *Turtle Island*. New York: New Directions.

———. 1980. *The Real Work: Interviews and Talks 1964–1979*. New York: New Directions.

———. 1983. "Wild, Sacred, Good Land." *Resurgence* 38:10–15.

Snyman, Jan W. 1986. "Man and Natural Resources: A Historical Perspective." *SUGIA Sprache und Geschite in Afrika* 7, no. 1:351–68.

Soles, J. S. 1991. "The Gournia Palace." *American Journal of Archaeology* 95:17–78.

Solheim, Wilhelm G. 1992. "Nusantao Traders beyond Southeast Asia." In *Early Metallurgy, Trade, and Urban Centers in Thailand and Southeast Asia*, edited by Ian Glover. Bangkok: White Lotus Press.

Spencer, Colin. 1993. *The Heretic's Feast: A History of Vegetarianism*. London: Fourth Estate Ltd.

Spondel, Walter M. 1986. "Kulturelle Moderuisierung durch antimoderuishshcen Protest: der lebensreformische Vegetarismus." *Koleur Zeitschrift fur Soziologie und Sozialpsychologie Sonderheft* 27:314–30.

Sponsel, L., and P. Natadecha. 1988. "Buddhism, Ecology, and Forests in Thailand: Past, Present, and Future." In *Changing Tropical Rainforests*, edited by J. Dargavel. Canberra, Aus.: Centre for Resources and Environmental Studies.

Spretnak, Charlene. 1990. "Ecofeminism: Our Roots and Flowering." In *Reweaving the World*, edited by I. Diamond. San Francisco: Sierra Club.

Springer, Robert. 1884. *Enkarpa: Culturgeschichte der Menscheit im Lichte der pythagoraischen Lehre*. Hannover: Schmorl & von Seefield.

Stark, Miriam. 1996. "Funan as Center or Funan as Periphery? Exploring State Formation in Mainland Southeast Asia." Paper presented at first East Asia Archaeological Network meetings. Honolulu, Hawaii.

Stein, Burton. 1998. *A History of India*. London: Blackwell.

Stern, M. 1991. "Early Roman Export Glass in India." In *Rome and India*, edited by V. Begley and L. DePuma. Madison: University of Wisconsin Press.

Storey, Glenn. 1997. "The Population of Ancient Rome." *Antiquity* 71: 966–78.

Su, Chung-Jen. 1967. "Places in Southeast Asia, the Middle East and Africa Visited by Cheng Ho and His Companions." In *Historical Archaeological and Linguistic Studies in Southeast Asia*, edited by F. S. Drake. Hong Kong: Hong Kong University Press.

Sweezy, Paul. 1976. "A Critique." In *The Transition from Feudalism to Capitalism*, edited by R. Hinton. London: New Left Books.

Szasz, Andrew. 1994. *Ecopopulism, Toxic Waste and the Movement for Environmental Justice*. Minneapolis: University of Minnesota Press.

Tadem, E. 1990. "Conflict over Land-Based Natural Resources in the ASEAN Countries." *In Conflict over Natural Resources in S.E. Asia and the Pacific*, edited by T. G. Lim and M. Valencia. Singapore: Oxford University Press.

Tahtinen, Unto. 1991. "Values, Non-Violence and Ecology: Two Approaches." *Journal of Dharma* 16, no. 3:211–17.

Tainter, Joseph A. 1988. *The Collapse of Complex Societies*. Cambridge: Cambridge University Press.

Tapar, B. K. 1993. "The Harappan Civilization: Some Reflections on Its Environments and Resources and Their Exploitation." In *Harappan Civilization: A Recent Perspective*, edited by G. Possehl. New Delhi: Oxford University Press.

Taylor, Charles. 1989. *Sources of the Self*. Cambridge: Harvard University Press.

———. 1992. *The Ethics of Authenticity*. Cambridge: Harvard University Press.

Taylor, Keith. 1992. "The Early Kingdoms." In *The Cambridge History of South East Asia*, vol. 1, edited by Nicholas Tarling. Cambridge: Cambridge University Press.

Tegengren, F. R. 1923–24. *The Iron Ores and Iron Industry of China*. Peking: Memoirs, Geological Survey of China Series A, no. 2.

Teng, Shu-Chu. 1927. "The Early History of Forestry in China." *Journal of Forestry* 25:564–71.

Thirgood, J. V. 1981. *Man and the Mediterranean Forest: A History of Resource Depletion*. London: Academic Press.

Thomas, K. 1983. *Man and the Natural World: A History of Modern Sensibility*. New York: Pantheon Books.

Thompson, William R. 1995. "Comparing the Evolution of Approaches to the Social Science History of the World System." Paper presented to Workshop on World System History: The Social Science of Long-Term Change. Lund University, Lund, Sweden, March 25–27.

Thomson, James T. 1988. "Deforestation and Desertification in Twentieth-Century Arid Sahelian Africa." In *World Deforestation in the Twentieth Century*, edited by John Richards and R. Tucker. Durham, N.C.: Duke University Press.

Thoreau, Henry David. 1971. *Walden*. Princeton: Princeton University Press.

——. 1972. *The Maine Woods*. Princeton: Princeton University Press.

——. 1982. *Thoreau in the Woods*. New York: Farrar, Straus, Giroux.

Thorley, J. 1971. "The Silk Trade between China and the Roman Empire at its Height Circa A.D. 90–130." *Greece and Rome* 18, no. 1:71–80.

Tibbetts, G. R. 1956. "PreIslamic Arabia and Southeast Asia." *Journal of the Malayan Branch of the Royal Asiatic Society* 29, no. 3:182–208.

——. 1957. "Early Muslim Traders in Southeast Asia." *Journal of the Malayan Branch of the Royal Asiatic Society* 30, no. 1:1–45.

Timmons, Roberts, and P. Grimes. 1997. "Carbon Intensity and Economic Development 1962–1991." *World Development* 25:191–98.

Tobias, Michael, ed. *Deep Ecology*. San Diego: Avant Books.

Torrance, Robert, ed. 1998. *Encompassing Nature: A Source Book on Nature and Culture from Ancient Times to the Modern World*. Washington, D.C.: Centerpoint Press.

Tosi, Mauricio. 1982. "The Harappan Civilization: Beyond the Indian SubContinent." In *Harappan Civilization: A Contemporary Perspective*, edited by G. Possehl. New Delhi: Oxford University Press.

Totman, Conrad. 1989. *The Green Archipelago*. Berkeley: University of California Press.

——. 1992. "Forest Products Trade in Pre-Industrial Japan." In *Changing Pacific Forests*, edited by John Dargavel. Durham, N.C.: Forest History Society.

Touraine, Alain. 1985a. "An Introduction to the Study of Social Movements." *Social Research* 52:749–88.

——. 1985b. *The Voice and the Eye*. New York: Cambridge University Press.

Toye, J. 1987. *Dilemmas of Development: Reflections on the Counter Revolution in Development Theory and Policy*. Oxford: Oxford University Press.

Toynbee, A. J. 1939. *A Study of History*. Vols. 4 and 5. Oxford: Oxford University Press.

Tuan, Yi-Fu. 1970. "Our Treatment of the Environment in Ideal and Actuality." *American Scientist* 58:244–49.

Tucker, Richard. 1988. "The Commercial Timber Economy under Two Colonial Regimes in Asia." In *Changing Tropical Forests*, edited by J. Dargavel. Canberra, Aus.: Centre for Resources and Environmental Studies.

Tucker, Richard, and J. F. Richards. 1983. *Global Deforestation and the Nineteenth-Century World Economy*. Durham: Duke University Press.

———. 1985. "The Global Economy and Forest Clearance in the Nineteenth Century." In *Environmental History*, edited by Kendall Bailes. New York: University Press of America.

Turner, B. L. 1985. "Issues Related to Subsistence and Environment among the Ancient Maya." In *Pre-Historic Lowland Maya Environment and Subsistence Economy*, edited by Mary Pohl. Cambridge: Harvard University Press.

Tzu, Lao. 1963. *Tao te ching*. Harmondsworth, Eng.: Penguin.

United Nations Conference on Environment and Development. 1992. *Earth Summit 1992*. London: Regency Press.

United Nations Economic and Social Commission for Asia and the Pacific. 1990. *State of the Environment for Asia and the Pacific*. Bangkok: ESCAP.

Utusan Konsumer. 1989. Newsletter of the Consumers Association of Penang. Penang, Malaysia.

Van Andel, T., and A. Demitrack. 1990. "Land Use and Soil Erosion in Prehistoric and Historical Greece." *Journal of Field Archaeology* 17, no. 4:379–96.

Van Andel, T., and C. Runnels. 1987. *Beyond the Acropolis: A Rural Greek Past*. Stanford: Stanford University Press.

———. 1988. "An Essay on the Emergence of Civilization in the Aegean World." *Antiquity* 62, no. 235:234–47.

Van Andel, T., C. Runnels, O. Pope. 1986. "Five Thousand Years of Land Use and Abuse in the Southern Argolid Greece." *Hesperia* 55, no. 1:103–28.

Van Andel, T., E. Zangger, and A. Demitrack. 1990. "Landscape Stability and Destabilization in the Prehistory of Greece." In *Man's Role in the Shaping of the Eastern Mediterranean Landscape*, edited by S. Bottema and G. Entjes-Nieborg. The Hague: Elsevier.

Van De Nieroop, Marc. 1992. "Wood in Old Babylonian Texts from Southern Babylonia." In *Trees and Timber in Mesopotamia*, edited by J. N. Postgate and M. A. Powell. Bulletin on Sumerian Agriculture 6. Cambridge: Cambridge University.

Van Vliet-Lanoe, P., M. Helluin, J. Pellerin, B. Valades. 1992. "Soil Erosion in Western Europe: From the Last Interglacial to the Present." In *Past and Present Soil Erosion*, edited by M. Bell and J. Boardman. Oxford: Oxbow.

Vermeule, Emily. 1960. "The Fall of the Mycenaean Empire." *Archaeology* 13, no. 1:66–75.

Vidale, Massimo. 1989. "Specialized Producers and Urban Elites: On the Role of Craft Industries in Mature Harappan Urban Contexts." In *Old Problems and New Perspectives in the Archaeology of South Asia*, edited by J. M. Kenoyer. Vol. 2 of Wisconsin Archaeological Reports.

Vishvanathan, V. 1997. "Politics of Indonesia's Forest Fires." *Economic and Political Weekly*, Oct. 11:2587–88.

Waddell, L. A. 1925. *The Indo-Sumerian Seals Deciphered*. London: Luzac Books.

Walberg, G. 1960. *Middle Minoan Provincial Pottery*. Mainz-am Rhein: Philipp von Zabern.

Wallerstein, Immanuel. 1974, 1980, 1988. *The Modern World-System*. Vols. 1–3. San Diego: Academic Press.

———. 1992a. "America in a Post-American World." *Humboldt Journal of Social Relations* 18, no. 1:217–23.

———. 1992b. "The West, Capitalism, and the Modern World System." *Review* 15, no. 4: 561–620.

———. 1995. "Hold the Tiller Firm: On Method and the Unit of Analysis." In *Civilizations and World Systems*, edited by Stephen Sanderson. Walnut Creek, Cal.: AltaMira Press.

Wang, Gung-Wu. 1958. "The Ninhai Trade: A Study of the Early History of Chinese Trade in Southeast Asia." *Journal of the Malayan Branch of the Royal Asiatic Society* 31, no. 2:1–135

———. 1989. *Sung-Yuan-Ming Relations with Southeast Asia: Some Comparisons*. Hong Kong: University of Hong Kong Press.

Warren, Peter M. 1985. "Minoan Palaces." *Scientific American* 253, no. 1:94–103.

Watson, William. 1992. "Pre-Han Communication from West China to Thailand." In *Early Metallurgy, Trade, and Urban Centres in Thailand and Southeast Asia*, edited by I. Glover. Bangkok: White Lotus Press.

Weber, Max. 1958. *The City*. New York: Free Press.

Weiner, M. 1991. "The Nature and Control of Minoan Foreign Trade." In *Bronze Age Trade in the Mediterranean*, edited by N. Gale. Goteborg: Paul Astroms Forlag.

Weiner, M. H. 1987. "Trade and Rule in Palatial Crete." In *The Function of the Minoan Economy*, edited by R. Hagg and N. Marinatos. Stockholm: Paul Astroms Forlag.

Weiss, B. 1982. "The Decline of Late Bronze Age Civilization as a Possible Response to Climate Change." *Climate Change* 4:173–98.

Wells, Bert, C. Runnels, and E. Zangger. 1993. "In the Shadow of Mycenae." *Archaeology* 46, no. 1:54–58.

Wertime, Theodore A. 1983. "The Furnace vs. the Goat: The Pyrotechnologie Industries and Mediterranean Deforestation in Antiquity." *Journal of Field Archaeology* 10:445–52.

Westra, Laura. 1997. "Aristotelian Roots of Ecology." In *Greeks and the Environment*, edited by L. Westra and T. Robinson. London: Rowman and Littlefield.

Wheatley, Paul. 1959. "Geographical Notes on Some Commodities Involved in the Sung Maritime Trade." *Journal of the Malayan Branch of the Royal Asiatic Society* 32, no. 2:1–140.

———. 1961. *The Golden Khersonese*. Kuala Lumpur: University of Malaya Press.

———. 1964a. *Desultory Remarks on Ancient History of the Malayan Peninsula*. Berkeley, Cal.: Institute of International Studies.

———. 1964b. *Impressions of the Malay Peninsula in Ancient Times*. Singapore: Eastern University Press.

———. 1971. *The Pivots of the Four Quarters*. Edinburgh: University Press.

———. 1980. *The Kings of the Mountain: An Indian Contribution to State Craft in Southeast Asia*. Kuala Lumpur: University of Malaya Press.

Wheeler, R. E. M. 1968. *The Indus Civilization*. Cambridge: Cambridge University Press.

Whitley, James. 1991. *Style and Society in Dark Age Greece: The Changing Face of a Pre-Literate Society 1100–700 B.C.* Cambridge: Cambridge University Press.

Whitmore, T. 1993. "Long-Term Population Change." In *The Earth as Transformed by Human Action*, edited by B. Turner, William C. Clark, Robert W. Kates, John Richards, Jessica Mathews, William B. Meyer. Cambridge: Cambridge University Press.

Whittaker, C. R. 1994. *Frontiers of the Roman Empire*. Baltimore: Johns Hopkins University Press.

Wilkinson, David. 1991. "Cores, Peripheries, and Civilizations." In *Core–Periphery Relations in Precapitalist Worlds*, edited by C. Chase-Dunn and T. Hall. Boulder: Westview Press.

———. 1995. " Civilizations Are World Systems!" In *Civilizations and World Systems*, edited by Stephen Sanderson. Walnut Creek, Cal.: AltaMira Press.

Willcox, G. H. 1974. "A History of Deforestation as Indicated by Charcoal Analysis of Four Sites in Eastern Anatolia." *Anatolian Studies* 24:116–33.

———. 1992. "Timber and Trees: Ancient Exploitation in the Middle East." In *Trees and Timber in Mesopotamia*, edited by J. N. Postgate and M. A. Powell. Bulletin on Sumerian Agriculture 6. Cambridge: Cambridge University.

Williams, M. 1990. "Forests." In *The Earth as Transformed by Human Action*, edited by B. L. Turner, William C. Clark, Robert W. Kates, John Richards, Jessica Mathews, William B. Meyer. Cambridge: Cambridge University Press.

———. 1997. "Imperialism and Deforestation." In *Ecology and Empire*, edited by Tom Griffiths and Robin Coles Libby. Seattle: University of Washington Press.

Winters, Robert K. 1974. *The Forest and Man*. New York: Vantage Press.

Wolters, O. W. 1967. *Early Indonesian Commerce*. Ithaca, N.Y.: Cornell University Press.

———. 1970. *History, Culture, and Region in Southeast Asian Perspectives*. Singapore: Institute of Southeast Asian Studies.

———. 1982. *History, Culture, and Region in Southeast Asian Perspectives*. Singapore: Institute of Southeast Asian Studies.

World Commission on Forests and Sustainable Development. 1999. *Our Forests, Our Future*. Cambridge: Cambridge University Press.

Wright, H. E. 1968. "Climatic Change in Mycenaen Greece." *Antiquity* 42:123–27.

Wright, Robin. 1989. "The Indus Valley and Mesopotamian Civilization: A Comparative View of Ceramic Technology." In *Old Problems and New Perspectives in the Archaeology of South Asia*, edited by J. M. Kenoyer. Vol. 2 of Wisconsin Archaeological Reports.

Wycherley, R. E. 1967. *How the Greeks Built Cities*. New York: Macmillan.

Yamamoto, T. 1981. "Chinese Activities in the Indian Ocean before the Coming of the Portuguese." *Diogenes* 3:19–34.

Yoffee, Norman. 1988. "The Collapse of Ancient Mesopotamian States and Civilization." In *The Collapse of Ancient States and Civilizations*, edited by N. Yoffee and G. Cowgill. Tucson: University of Arizona Press.

Yorimitsu, Ryozo. 1992. "Thriving Housing Construction in Japan and the Factors Contributing to It." In *The Current State of Japanese Forestry*. Tokyo: The Japanese Forest Economic Society.

Zangger, E. 1993. "Neolithic to Present Soil Erosion in Greece." In *Past and Present Soil Erosion*, edited by M. Bell and J. Boardman. Oxford: Oxford University Press.

Zimmerman, Francis. 1987. *The Jungle and the Aroma of Meats: An Ecological Theme in Hindu Medicine*. Berkeley: University of California Press.

Index

Page numbers in italics refer to tables or figures in text.

About the Author

Sing C. Chew is Professor and Chairperson of the Department of Sociology at Humboldt State University in Arcata, California. Prior to his current appointment, he was the Associate Director in the Office of Vice-President (Resources), International Development Research Centre, Ottawa, Canada. He has been a Research Fellow at the Institute of Southeast Asian Studies (ISEAS) in Singapore, and, recently, Guest Researcher at the Human Ecology Division, Lund University, Lund, Sweden, where this book was completed. His most recent book is a coedited volume: *The Underdevelopment of Development: Essays in Honor of Andre Gunder Frank.*